Energy Audit and Management

This book describes the energy management concepts, energy audit principles, resource efficiency, and other energy conservation opportunities involved in different sectors across varied industries. Real-time case studies from various large industrial sectors, like cement, paper and pulp, refineries, manufacturing, garments and textile processing, power plants, and other MSME industrial sectors with cross functional energy conservation opportunities, are included. It also describes the future scope of energy auditing and management including IoT and data analytics. It also helps to gather the energy generated and utilization, energy conservation, and other process related data.

Features:

- Provides entire coverage of energy management and audit concepts.
- Explores energy audit methodologies and energy saving initiatives.
- Incorporates current technologies like machine learning, IoT, data analytics in energy audit for reliability improvement.
- Includes case studies covering detailed energy saving calculations with investment pay back calculations.

This book is aimed at researchers, professionals, and graduate students in electrical engineering, power systems, energy systems, and renewable energy.

Energy Audit and Management

Concept, Methodologies, Procedures, and Case Studies

Edited by L. Ashok Kumar and Gokul Ganesan

CRC Press
Taylor & Francis Group
Boca Raton London New York

CRC Press is an imprint of the
Taylor & Francis Group, an **Informa** business

Designed cover image: Shutterstock

First edition published 2023
by CRC Press
6000 Broken Sound Parkway NW, Suite 300, Boca Raton, FL 33487–2742

and by CRC Press
4 Park Square, Milton Park, Abingdon, Oxon, OX14 4RN

CRC Press is an imprint of Taylor & Francis Group, LLC

ISBN: 978-1-032-06779-7 (hbk)
ISBN: 978-1-032-06781-0 (pbk)
ISBN: 978-1-003-20381-0 (ebk)

DOI: 10.1201/9781003203810

Typeset in Times
by Apex CoVantage, LLC

Contents

Editor Biographies

Dr. L. Ashok Kumar was Postdoctoral Research Fellow from San Diego State University, California. He is a recipient of the BHAVAN fellowship from the Indo-US Science and Technology Forum and SYST Fellowship from DST, Government of India. His current research focuses on integration of Renewable Energy Systems in the Smart Grid and Wearable Electronics. He has three years of industrial experience and 19 years of academic and research experience. He has published 167 technical papers in international and national journals and presented 157 papers at national and international conferences. He has completed 26 Government of India–funded projects, and currently seven projects are in progress. His PhD work on wearable electronics earned him a National Award from ISTE, and he has received 24 awards on the national level. Ashok Kumar has nine patents to his credit. He has guided 92 graduate and postgraduate projects. He is a member and in prestigious positions in various national forums. He has visited many countries for institute industry collaboration and as a keynote speaker. He completed his graduate program in Electrical and Electronics Engineering from University of Madras and his postgraduate from PSG College of Technology, India, and Masters in Business Administration from IGNOU, New Delhi. After completion of his graduate degree, he joined as project engineer for Serval Paper Boards Ltd., Coimbatore (now ITC Unit, Kovai). Presently he is working as Professor in the Department of EEE, PSG College of Technology and also doing research work in wearable electronics, smart grid, solar PV, and wind energy systems. He is also a Certified Chartered Engineer and BSI Certified ISO 500001 2008 Lead Auditor. He holds prestigious positions in various national and international forums, and he is a Fellow Member in IET (UK), Fellow Member in IETE, Fellow Member in IE, Senior Member in IEEE, and Chairman of IAEMP Coimbatore Chapter.

Mr. Gokul Ganesan was Product Sustainability and Digitalization Professional with a postgraduate degree in Energy Engineering from PSG College of Technology, Coimbatore. He has completed his undergraduation work in Electrical and Electronics Engineering from Dr. N.G.P. Institute of technology, Coimbatore. He is Executive Council member of Indian Association of Energy Management Professionals (IAEMP), Central Council and Coimbatore Chapter. His current research focuses on Improving Product Sustainability, Industry 4.0, Energy Management, Energy Analytics, Smart Grids, and Renewable Energy. He has six years of research and industrial experience. He has conducted more than 75 Detailed Resource Efficiency Audits in various MSMEs (garments, spinning mills, textile process industries, and foundries) and ten Detailed Energy Audits in the cement industries and thermal power plants. He has assisted more than 80 MSMEs in developing energy and resource efficiency projects, technology upgradation, process improvement, and project implementation and reduction in their Specific Energy Consumption (SEC). He has trained 80+ MSMEs in Energy Efficiency and Best Practices through various training programs and awareness campaigns over social media. He has conducted several hundreds of surveys if different sectors to identify the financial requirement toward energy efficiency. He has conducted 50+ webinars and workshops among school, college, and industry professionals on Industry 4.0, Energy Audit, Energy Management, and Energy Analytics representing IAEMP Coimbatore Chapter. He has represented IAEMP on Mentoring 30+ postgraduate students and 25+ undergraduates in energy management, energy audit, and green buildings.

Contributors

Dr. A. Amudha
Professor, Department of EEE
Karpagam University
Coimbatore, India

Dr. L. Ashok Kumar
Professor and Associate HoD,
    Department of EEE
PSG College of Technology
Coimbatore, India

Dr. L. Chitra
Assistant Professor (SG)/EEE
Dr. Mahalingam College of
    Engineering & Technology
Pollachi, India

Ms. Dhiksha Mohan
Energy Research Analyst
International Society for Energy and
    Sustainability Research (ISESR)
Coimbatore, India

Mr. Gokul Ganesan
Executive Council Member
Indian Association of Energy
    Management Professionals (IAEMP)
Coimbatore, India

Mr. Gokul Rajendran
Member, Indian Association of Energy
    Management Professionals (IAEMP)
Coimbatore, India

Dr. J. Jayakumar
Professor, Department of Electrical and
    Electronics
Karunya University
Coimbatore, India

Dr. Josephine Rathinadurai Louis
Department of EEE
National Institute of Technology
Tiruchirappalli, India

Mr. Kishen Singh
Member, Indian Association of Energy
    Management Professionals (IAEMP)
Coimbatore, India

Ms. Krishna Rubigha
CEO (India) International Society for
    Energy and Sustainability Research
    (ISESR)
Coimbatore, India

Mr. D. Mohankumar
Assistant Professor, Department of
    Mechanical Engineering
KPR Institute of Engineering and
    Technology
Coimbatore, India

Rachaputi Bhanu Prakash
Department of EEE
National Institute of Technology
Tiruchirappalli, India

Mr. Sivakumar
Cluster Leader, GEF-UNIDO-BEE
    Project
COINDIA
Coimbatore, India

Ms. Y. Uma Maheswari
PhD Scholar
Karpagam Academy of Higher
    Education
Coimbatore, India

Preface

Sustainability is one of the major factors burgeoning in today's world. Sustainability, energy management, and energy conservation are the key objectives and goals of every organization. It is an essential practice to conserve energy in order to reduce the global warming. Conventional energy audit methods are changing day by day in this digital world. The future of energy audits or energy management will be focused more and more on forecasting energy demand and generating power as needed. This concept involves technologies such as data analytics, machine learning, Industrial Internet of Things, and cybersecurity. Updating the skills along with the market trends is an important factor, but at the same time understanding the fundamentals and basic concepts of energy conservation and management will help in implementing these advanced technologies. Industries are now focusing more toward the cross functional concepts along with real-time case studies to implement in their organization.

Acknowledgments

The authors are always thankful to the Almighty for their perseverance and achievements.

The authors owe their gratitude to Shri L. Gopalakrishnan, Managing Trustee, PSG Institutions, and to Dr. K. Prakasan, Principal in Charge, PSG College of Technology, Coimbatore, India, for their wholehearted cooperation and great encouragement in this successful endeavor.

The authors express a special thanks to the contributors of this book on various chapters Dr. Jayaram, Mr. Kishen Singh, Mr. Gokul Rajendran, Mr. Sivakumar, Ms. Y. Uma Maheswari, Ms. A. Amutha, Josephine Rathinadurai Louis, Mr. Mohan, Ms. Krishna Rubiga, Ms. Dhiksha Mohan, Dr. L. Chitra.

Dr. L. Ashok Kumar would like to take this opportunity to acknowledge those people who helped in completing this book. He is thankful to all his research scholars and students who are doing their projects and research work with him. But the writing of this book is possible mainly due to the support of his family members, parents, and sisters. Most importantly, he is very grateful to his wife, Y. Uma Maheswari, for her constant support during writing. Without her, all these things would not be possible. He would like to express his special gratitude to his daughter, A. K. Sangamithra, for her smiling face and support, and would like to dedicate this work to her.

Mr. Gokul Ganesan owes thanks to his parents Mr. S. Ganesan and Mrs. G. Maruthamanickam for their continued encouragement for completing the book. Mr. Gokul Ganesan would like to thank his wife, Mrs. G. Suganya and his sister Ms. G. Sandhya for their unconditional support and time whenever needed during the writing of the book. He would also like to extend his gratitude to his friends Mr. Praveen Kumar and K. Kasi Aswin for their technical support and advice.

The authors wish to thank all their friends, colleagues, and research assistants who have been with them in all their endeavors with their excellent, unforgettable help and assistance in the successful execution of the work.

Acknowledgments

Organization of the Book

Chapter 1 describes the basics of energy, various forms of energy, energy audit concepts and its methodology, the different types of energy audit, and the need for an audit in a specific industry. This chapter also discusses the need for Energy Auditors, their roles and responsibilities.

Chapter 2 discusses various energy conservation techniques in thermal systems. The chapter explains in detail the saving opportunities in fuel utilization, steam distribution, furnaces, cooling towers, and several other factors in a thermal system. Energy saving calculations for all the opportunities have been explained.

Chapter 3 gives an insight to the basics of HVAC technologies and refrigeration systems, and the multiple types of HVAC and refrigeration systems. The chapter also describes various energy conservation opportunities in HVAC systems along with real-time implemented case studies with detailed technocommercial calculations.

Chapter 4 provides a detailed explanation of industrial furnaces and different types of furnaces used in industries and their applications. This chapter explains the energy saving techniques practiced in different industries in furnaces and the energy saving calculations.

Chapter 5 explains the energy conservation methods applied in electrical systems. The chapter describes in detail the energy conservation opportunities in electrical systems, motors, lightings, transformers, and other areas of electrical systems inside the factory.

In Chapter 6, the effect of electromagnetic interferences (EMI) in electrical systems are described in detail. The effect of EMI due to switching devices in converters and drives is discussed, as well as how the EMI effect changes the efficiency of the system.

Chapter 7 presents an analysis on efficient energy conversion techniques for PV-fed induction motor drive in irrigation applications. This chapter explains the effect of the integration of renewable energy into the equipment.

In Chapter 8, the essential ideas behind intelligent universal transformers (IUT) introduced. The chapter provides an insight into the applications of IUT in several industries and how the intelligent universal transformers can help in energy conservation.

In Chapter 9, energy management systems, the Internet of Things (IoT), Industry 4.0 (I4.0), and the need for the Internet of Things and I4.0 in the field of energy management are explained.

In Chapter 10, we discuss the recent advances in nonintrusive load monitoring (NILM) for energy management. Also explained is the importance of energy management and data acquisition in the field of energy sector to achieve energy optimization and energy conservation.

Chapter 11 explains the goals set by an educational institution toward energy conservation in their campus and the achievements. A detailed technocommercial explanation on a real-time implemented energy saving proposal is explained.

Chapter 12 describes the optimization techniques for optimal power flow—a precursor. The methods of optimization in power system operation scheduling, analysis, and management of energy with the help of optimal power flow is also explained.

1 Fundamentals of Energy Management and Energy Conservation

L. Ashok Kumar

CONTENTS

DOI: 10.1201/9781003203810-1

LEARNING OUTCOMES

At the end of this chapter, the reader will be able to understand:

- The basics of energy, energy audit, concept and methodology.
- The need for an energy audit and its types.
- Multiple forms of energy sources and reduction methods.
- The need for Energy Auditors, their roles and responsibilities.

1.1 INTRODUCTION

Slogans like "Switch off fan, lights, while leaving" and "One unit of energy saved equals 2 units of energy generated" express the real objective of this book. The primary prerequisite is that "common sense in legal matters is a great asset to the Energy Auditors." Someone once raised this question: "mass and energy balance—is it really done by Auditors?" The mass and energy balance is not required for all cases. It is a costly affair. In 1995, an exhaustive case study was carried out for a company in the cement industry. At that time, the company itself paid Rs. 8 lakhs for the Auditor. About 20 engineers were split into five teams and took 45 days to complete an energy audit. Today, that exercise would cost a minimum of Rs. 50 lakhs. Inevitably, the question is demand is made to "show the report of the companies where you have done mass energy balance." The Energy Auditors, always serious-minded people, are interested only in their calculations and their ways and means to save energy. In the process, they fall into the trap of promising company executives an audit after a brief visit to their factory. At the end of day, they wind up having a free lunch at the cost of valuable information that they divulge during the discussion. The moral of the story is, "Be shrewd at business to be a successful Auditor." Engineers always want to maintain their originality and are pleased when someone appreciates their work. For as long as we fail to press the alarm button, there is no solution for the vexing energy conservation situation. People, i.e., consumers, the public, must wake up and hit the alarm bells. There is no use in switching off the lights for an hour in order to make people aware of the seriousness of the problem. Everybody has to save energy in their homes and offices.

It is a fact that the certified Auditor and the experienced engineer have good theoretical knowledge of energy conservation principles and methods. But technical skills alone are not sufficient. Certain business skills are also required for advancement in the trade. Nobody will tell you the marketing secret, including your best friend. The financial rewards are spectacular for those who master the tricks of the trade and its secrets. Do not play the blame game telling the age-old story that there is no direction from the government and hence there is no demand for the Auditors. This is true only to a certain extent. Even now, some Auditors are doing well. Just know that many young engineers have hit the market in a big way, only to fail miserably, losing everything they had.

A good Energy Auditor should be able to approximately estimate the percentage saving within 2 hours of discussion with plant engineers, going round the plant, and looking at the data collected. The first job of the Energy Auditor is to identify major energy saving proposals with or without investment. It is easy to suggest replacing the old equipment with automated modern energy efficient equipment. Energy Auditors should not scare away management by talking about impractical suggestions involving huge investments. Some Energy Auditors are more interested in pushing certain companies' products for ulterior motives. Hence the credibility of auditors at large is at stake. The most important task of an Energy Auditor is to conduct the training program for the energy managers/engineers of the company. Training is a neglected aspect in most organizations even though everybody talks outwardly a lot about the importance of human resources development.

Energy Auditors should have high integrity. The measurement of plant parameters is very important and helps in the decision-making process. For example, in the cement industry, the measurement of fan flow and the calculation of efficiency of the fan at plant operating conditions play a key role. The fans in the cement industry consume around 30% of total energy consumption. A single measurement of the parameters is not sufficient. The locations of sampling points have to be decided as per standards to ensure accuracy of reading.

Big industrial houses would have already done energy auditing by internal and external auditors. What they require is only fine-tuning of the energy levels by establishing proper dynamic norms for each and every section and achieving further reductions in energy consumption. Energy Auditors may do three types of audits:

1. **Energy audit for the industry and building:** The Energy Auditor is familiar with the E.C. Act 2001, which deals with this.
2. **Energy audit for the individual:** The individual should have sufficient energy to withstand the stresses and strains in business and life. This audit tells how to develop one's own energy level so that the person remains hale and healthy.
3. **Energy audit for the environment:** The environment we live in should have positive vibes, so that one feels pleasant whether in one's own home, office, or some other place.

1.2 INTRODUCTION TO ENERGY AUDITING

Energy conservation is a critical measure for environmental protection and additionally for reducing exchange outflow, which is used to acquire polluting fossil fuels, primarily petrol. Energy conservation is critical in homes and tertiary structures, where electromechanical equipment and domestic appliances consume 30% of the country's total energy consumption, growing at a rate of 4% per year since the mid-1970s. Additionally, the operation of energy systems placed in buildings contributes 40% of total CO_2 emissions to the atmosphere, a gas that contributes to the planet's greenhouse effect. While industrial consumption has been gradually declining in recent years (mostly due to a drop in energy-intensive industrial sectors), its contribution to total energy consumption in the country is significant.

The energy crisis is the one of the crucial problems faced by all the countries in the world due to depletion in natural resources used for energy generation and the huge investment for generating energy from alternate resources. A viable and immediate solution in this juncture is the energy conservation as cited by the slogan, "Energy conserved is energy generated." To conserve energy, an energy balance sheet has to be prepared by doing energy auditing, which is an accounting of electrical energy in every stage of energy transformation from generation to utilization. Any prospective electricity board, electricity-generating company, captive power plant owners, transmission or/and distribution companies, industries, consumers can do energy accounting studies and reduce electrical energy losses in their systems.

The reduction in transmission and distribution (T&D) losses would improve the efficiency, performance, reliability, and stability of the electrical system at minimum cost. This results in energy savings directly and improvement in voltage profile indirectly, in addition benefiting the state electricity boards twice by increases in revenue by means of additional energy sales with the saved energy without investing in infrastructure for new power generating stations. In addition, this results in better consumer satisfaction by ensuring quality power supply. For any prospective electricity user, reduction in loss leads to reduction in consumption and electricity charges.

As electricity is an essential commodity without which no one can live, electricity demand due to increases in population and civilization is also on the rise. All the nations of this world are trying to bridge the gap between power generation and demand. But they face an energy crisis due to (1) depletion in natural resources such as coal, oil, gas, etc., (2) the unavailability of new major hydro potential resources for exploration, (3) huge investment costs in nuclear resources, and (4) nonavailability of major, nonconventional, reliable, stable, and perennial electrical energy generating systems. Hence energy conservation is the need of the day as cited by the slogan, "Energy conserved is energy generated." Any prospective electricity board, electricity-generating company, captive power plant owners' transmission or/and distribution company, industry, or consumer can do energy accounting studies and reduce electrical energy losses in their systems.

The diachronic route of energy indexes/indicators is the outcome of our country's rapid improvement, along with generally low-quality construction/building practices on the shell and installations of structures. These two parameters are inextricably linked by the absence of an integrated decretory framework of incentives and energy-efficient building regulations, as well as a viable national energy saving program aimed at improving construction/building quality and increasing user awareness of energy issues.

1.2.1 Energy Audit—What? Why? How?

Energy audit is a crucial step because it provides a means for identifying the areas of energy use and for establishing the priorities. The effectiveness of the energy utilization varies from industry to industry because of the diversity of the various processes. Hence a tailor-made energy conversation scheme should be prepared to meet the specific needs and to match it with the overall goal of the company. A well planned organized energy conservation program is a must for success. The energy inputs—both electricity and fuel—are the essential parts of any manufacturing process and often form a significant part of the expenditure of the plant. Thus any saving in energy directly adds to the profits of the company. The cost of energy is escalating and will continue to do so in the near future. Also, availability itself is uncertain and unsteady.

1.2.2 Purpose of Energy Audit

The purpose of an audit program is to systematically identify opportunities for saving energy in the plant and to realize the benefits of cost reduction. It also involves identifying the loss of energy, retrofitting and modification, alternate energy sources, etc. Our experience shows that 5–25% of energy can be saved in any plant by carrying out systematic and scientific energy auditing. The approach should be to study, observe, analyze, measure, and quantify useful energy and wastages. Also, the Energy Auditor should identify practical opportunities for saving for as long as it is economically feasible and practically possible. The realistic targets should be set for long-term energy saving.

1.2.3 CLASSIFICATION OF ENERGY AUDIT

The type of energy audit has to be performed depends on three factors:

1. Type of industry
2. Desirability of cost reduction
3. Depth of final audit needed

Depending on these factors, energy audits can be categorized in two types:

1. Preliminary audit
2. Detailed audit

The preliminary audit identifies the immediate need of the plant, such as:

1. Establishing the energy consumption of the plant.
2. Estimating the savings.
3. Identifying areas which requires immediate attention.
4. Identifying areas where detailed study is required.

A detailed audit consists of three phases:

1. Phase I: preaudit phase
2. Phase II: audit phase
3. Phase III: postaudit phase

1.2.4 SCOPE OF ENERGY AUDIT

The energy audit for the factory covers the following:

DETAILED ENERGY AUDIT

1. Professional studies for improving energy efficiency
2. Study of motors and their loadings to identify the inefficient and oversized motors
3. Study of air conditioning and humidification system
4. Power factor correction at load end and study of line losses
5. Lighting study for energy saving in lighting systems
6. Optimization of the cooling water system, water conservation
7. Optimization of compressed air system/steam system/refrigeration

8. Study of DG sets, etc.
9. Study of the effectiveness of utilities like water treatment plants, effluent treatment plants, air systems
10. Instruments for energy conservation
11. Study of pumps/humidification system
12. Development of management implementation system for energy conservation program
13. Fixation of energy norms
14. Training of manpower on energy auditing

1.2.5 Methodology Adopted for Energy Audit and Energy Conservation Technology

The Energy Auditor should look for the opportunities as per the following format:

1. Housekeeping
   a. Management information system
   b. Preventive maintenance schedule
   c. Training
   d. Steam leak/traps
   e. Air leak/misuse
   f. Insulation
   g. Idle running
   h. Optimization of process parameters
2. Retrofitting
3. Projects
   a. Cogenerating/wind mills
   b. New technology/technology upgradation
4. Energy conservation at macrolevel
   1. Increase in production
   2. Fine-tuning the energy levels
   3. Upgrade of technology
   4. Adoption of energy efficient technologies in industries

1.2.6 Benefits of Improved Energy Efficiency in Buildings and Industrial Facilities

Energy efficiency (EE) measures used in the construction and industrial sectors may result in three separate benefits:

1. Financial benefits that result in a decrease in an organization's operational costs or in an increase in its earnings. These must be weighed against the costs associated with implementing the EE measures.

- Operational benefits that assist in the administration of an industrial site or building by increasing the inhabitants' comfort, safety, and productivity, or by otherwise improving the site's overall operation

- Environmental benefits, including reduced CO_2 or other greenhouse gas emissions, decreased national energy consumption, and conservation of natural resources
- The gradual accrual of each advantage and its cumulative effect (The primary benefits may be instantly available through no-cost solutions, or they may need some time before a return on investment is realized. Others may be realized only through the implementation of long-term strategies.)

1.2.7 TYPES OF ENERGY AUDITS (WALK-THROUGH AND DETAILED ENERGY AUDITS)

Energy audits are classified in two forms based on the level of detail acquired during the collection of data: walk-through audits and extended audits.

On the basis of energy bills-invoices and a brief on-site inspection, walk-through energy audits determine a site's energy use and associated expenses. Housekeeping or/and low-capital-investment energy saving opportunities with an immediate economic return are identified, and a cost–benefit analysis of other energy saving opportunities requiring frequently substantial capital is performed.

Detailed energy audits require more extensive site data recording and analysis. Diverse aspects that affect various end uses (heating, cooling, different processes, lighting, etc.) are outlined and analyzed, and the energy consumption is disaggregated in each end use (e.g., production or services capacity, climatic conditions, raw material data, etc.). The advantages and costs for every possible energy saving initiative are quantified. The report also includes a list of possible investments with details on the information needed to acquire and process it.

1.2.8 OVERVIEW OF THE ENERGY AUDITS GENERAL PROCEDURE

Energy auditing (EA) uses building audits and energy installation checks to help figure out how to save energy. After entry of the data, the audits must go through three separate stages: registration, diagnosis, and treatment.

First Stage: Scheduling an Energy Audit—Primary Data Collection and Basic Energy Analysis

At this stage, statistics and information are gathered about the current/historical energy profile of each building/unit, as well as the unit's construction and use. This data/information can be gathered using a systematic and brief questionnaire (see Annex A), which is completed following the Energy Auditor's initial visit with the building/unit manager who requested the energy audit. The questionnaire requires information from the building/technical units and administrative managers, as well as existing pertinent data (fuel bills/invoices, technical drawings, archival studies and catalogues, recorded measurements and readings, and so on).

The preliminary analysis of the collected data should result in the identification of the annual trend and monthly fluctuation/variation of the audited building's/total unit's energy consumption and cost, which comprise the energy profile of the building/unit. The initial energy data obtained should also result in a first approximation of the building/energy unit's consumption allocation in each region and subsystem. This is a means of expressing the building's/energy unit's balance. At the conclusion of this stage, the Energy Auditor can compile/create an initial catalogue/list of potential energy saving actions/activities for each building/unit, taking into consideration any owner-imposed exclusions.

Second Stage: Brief Walk-through Energy Audit

This stage involves conducting a qualitative assessment of the building shell and electromechanical equipment and compiling the findings in a prescribed format. This data collection, together with immediate sample measurements, enables the allocation of energy use and hence the determination of the building's/energy unit's balance.

This technique, when combined with the previous stage's activities, results in a final determination of energy savings potential through the employment of tying-up measures and simple, low-cost measures/actions that do not require economic payback analysis via applicable energy studies. Additionally, it results in the identification of energy saving potential in specific locations and systems, which can be further examined in a subsequent stage by specialists/consultants or by building administration employees, as appropriate. These possible energy saving measures must be classified into three categories based on their energy saving potential for the specific structure (high, medium, low).

Third Stage: In-site Thorough Energy Audit

It entails data gathering (from on-site measurements) and processing, as well as a complete analysis of the building's/installed unit's energy systems, which enables the creation of a comprehensive energy data. Additionally, this technique enables a sound technological and economic evaluation of one or more energy saving approaches, requiring moderate to high investments in specific systems, following a relevant investigation.

The energy audit procedure concludes with the presentation of all energy saving ideas in the form of a summary technoeconomic report prepared by the Energy Auditor and delivered to the building/unit manager.

1.2.9 TYPICAL TOOLS AND TIME FRAMES FOR ENERGY AUDITS IN VARIOUS APPLICATIONS

In general, the typical requirements for conducting energy audits are as follows: staff with relevant knowledge/skills adequate time to perform the tasks involved, technical equipment for any measurements, financial assistance in the preceding areas and for implementing the recommendations, and technical and operational information on buildings/plant/services.

The time required to conduct an energy audit varies according to the availability of energy data, the site's size, and the complexity of the systems. A walk-through audit may take only a few hours for a modest site with clearly accessible content. On more sophisticated systems, one may easily spend a week or more merely analyzing invoices and records.

There are no hard-and-fast rules for determining how much time should be spent on a site survey; it depends on the site's complexity, the availability of resources, and the amount of money that can be justified. An estimate can be constructed by examining the various components to be analyzed. The largest of these locations may require the equivalent of one person-year to audit thoroughly or preferably a small team to complete the audit in a shorter amount of time. Within a day, one person might conduct a short audit of a small structure.

Allow time for those conducting the audit as well as those helping in other ways. This may include providing information or acting as an escort. Even with outside aid, some involvement from internal staff is always required. The greater the collaboration is, the more accurate the energy audits will be. Therefore, employees should be encouraged to make a constructive contribution.

In terms of instrumentation, it's worth noting that measurement lays the groundwork for comprehending the energy flow on-site. Metering and equipment enable a quantitative analysis of energy use and service quality. With expert application and experience in interpreting data, it is possible that significantly more knowledge will be gathered than from observation alone. The need for calibrating instruments appropriately in order to receive reliable data is emphasized.

Temporary test equipment is widely available for purchase or rental for the majority of applications where a clear need for reliably measured data is determined. Such equipment should be utilized only to the extent that is absolutely essential to get sound findings. A well conducted test avoids the generation of excessive data with a high number of readings or a prolonged recording period.

The most frequently necessary measures are those of the environment, power, air handling, piped services, and the boiler plant. A succinct audit may need to use some instrumentation. Comprehensive audits are often expected to involve the measurement of significant energy flows and an assessment of the operation of large plants. For full assessment, accurate measurements of building areas and volumes are also required.

1.3 PLANNING AN ENERGY AUDIT

1.3.1 Scope and Needs of an Energy Audit

An energy audit is a broad phrase referring to a systematic approach that tries to gain a thorough understanding of a building's or industrial plant's energy usage profile. Additionally, it tries to find and scale up energy saving solutions for the unit. Energy audits are critical for implementing energy conservation measures and ensuring that energy management targets are met.

In an energy audit, the primary objective is to save energy; the focus is on energy consumption and possible savings. Other factors may be considered (technical condition, environment), but the primary focus is on energy savings, and the audit generates reports on energy saving strategies. The audit work may include all energy-related features of a site or only a subset of the site's energy-related components (systems, equipment) (i.e., horizontal audit). In other circumstances, the entire process may be referred to by another name (such as energy labeling or energy assessment), although the activity fits the same criteria as an energy audit. Energy auditing is not a continuous process but should be conducted on a regular basis.

1.3.2 Design Criteria for an Energy Audit

To ensure that an effective energy audit can be developed at a reasonable cost and to account for the range of audit types, the entire procedure should be organized around particular criteria. The following aspects should be mentioned or considered during the planning stage of an energy audit:

- **Staff involvement:** Ideally, the audit should be directed by a person with managerial or board-level authority to lend legitimacy to the audit and its findings. Whether outside aid is utilized is largely determined by the complexity of the site, its buildings and systems, and the availability of competent personnel.
- **Site or building border:** Typically, a single building, such as an office block, poses minimal issues about the audit's boundary. It is frequently desirable to identify each individual building to be included on a site with many buildings, especially if the buildings are of varying construction and usage. Additionally, it is critical to note any facility or department that will be excluded from the audit for a stated reason.
- **Scope of the energy audit and the level of detail:** This should be determined by the available resources and the projected energy savings opportunities.
- **Audit timeliness:** A well-timed audit will yield the best findings. Seasonal and other planned activities should be incorporated into programming.
- **Access to the location:** Checking staff and working procedures may be restricted. Department heads and security personnel should be informed of the program and urged to cooperate in order for the audit to function smoothly.
- **Requirements for reporting:** Reporting processes must be considered early on. Take note that the time spent assessing findings and creating a final report is typically at least as extensive as the time spent on-site.

1.3.3 Preliminary Energy Considerations

Energy consumption and output data, where applicable, are required for all but the simplest energy audits. To this aim, data collection and the monitoring of consumption and output should begin immediately upon consideration of auditing. The longer the data collection time, the more beneficial the energy auditing operation will be. Even if complete records are available, such as monthly electricity bills, independent weekly or daily meter readings can be extremely helpful in analyzing energy use, loss, or savings. Additionally, when performing an energy balance, more comprehensive consumption data aid in quantifying individual energy flows and improve the balance's accuracy.

In the initial step of an energy audit of a facility or unit, it is important to collect preliminary data on the building's or unit's energy consumption behavior. Thus, it is necessary to complete the following information:

- General information about the building (type of building, year of construction, type of use and provision of services, ownership status, authorized representative, possible shell renovations-extensions to the building's installations, building area and volume, number of users, products and related service support equipment, operating status, typical floor plan layout)
- Data on energy use and invoicing for the previous five years (annual evolution of fuel and electricity consumption, monthly variation in consumption within the year under investigation)
- Energy management status and any energy saving measures implemented

Additionally, supporting data must be gathered:

- Energy bills and invoices (electricity, fuels) for the audit period, as well as the preceding four years (or/and the next four years)
- Plans and studies for the structure's electromechanical energy systems
- The basic apparatus/appliances' construction/structural and operating features
- Climate data for the auditing era
- Archived recordings including readings from current recorders or theoretical estimates of the building's/energy unit's consumption levels

The form is completed and supporting data are collected by the building's manager in collaboration with the Energy Auditor. Additionally, all data and information on the building/unit can be placed into an appropriate database for future retrieval, which is beneficial not only for the building/unit in question but also for comparison purposes with other similar buildings and installations.

Proposed Tasks of an Energy Audit

The suggested work assignments for an energy audit should be sufficiently adaptable to allow for changes that may occur throughout the course of the relevant work, based on the acquired data and in order to maximize the utilization of available resources. However, it should be noted that depending on the type of energy audit being conducted, the required work duties that the appropriate auditor must complete also vary. Thus an energy audit's general responsibilities may include the following:

- The audit's objective, objectives, and requirements
- The direction of the complex's units, systems, and/or buildings to be audited
- The description of the audit's tasks and stages
- Establishment of the standards and procedures to be followed during the audit (These requirements, as well as the many regulations cited, must be clearly mentioned.)
- The assignment of departments or other personnel inside the audited organization who will collaborate with the Auditor during the energy audit's completion

- The identification of further research, databases, and other pertinent sources to be used to gather reference data
- Analysis of the time required to complete the audit's duties
- Specification of the members of the energy audit team
- Possible requests for confidentiality

1.4 ENERGY AUDIT CONCEPT

Energy audit is a critical step because it provides a means for identifying the areas of energy use and for establishing priorities. The effectiveness of the energy utilization varies from industry to industry due to the diversity of the various processes and activities being performed. Hence a tailor-made energy conservation scheme should be prepared to meet the specific needs and to match it with the overall goals of the company. A well planned and organized energy conservation program is a must for success. The energy inputs—both electricity and fuel—are the essential parts of any manufacturing process and often form a significant part of the expenditure of the plant. Thus any saving in energy directly adds to the profits of the company. The cost of energy is escalating and will continue to do so in the near future. Also, availability itself is uncertain and unsteady.

1.4.1 ENERGY PARAMETERS ESTIMATION AND MEASUREMENTS

One of the primary objectives of an energy audit, particularly one of the thorough (detailed) variety, is to configure energy standards for reference consumption, reference specific consumption, and reference efficiency for individual installations and devices. With the aid of such standards, it is possible to assess energy consumption before and after the implementation of energy saving measures. Energy standards must be highly sensitive to critical characteristics such as production volume, raw material quality and composition, operating schedule, and ambient temperature. As a result, it is important to quantify and evaluate a set of parameters classified as follows:

- **Energy input for the unit's end usage, such as electricity and fuels:** It is vital to incorporate measures of the heating value, humidity, ash, stable carbon, and volatile compounds when dealing with solid fuels. In the case of renewable energy sources (RES), measurements extend to natural parameters that define the RE Source's intensity (e.g., annual mean wind speed).
- **Energy flow, conversion, and losses:** This must be known in of the various manufacturing and construction facilities, including steam, hot water, electricity, thermal radiation, and compressed air.
- **Energy-related operational circumstances:** Variations and mean values of temperature, humidity, pressure, fluid velocity, and luminance levels, exist in production installations and building spaces. Additionally, operating hours and idle/shutoff frequencies are measured.
- **When raw materials, intermediate products, and finished products are directly tied to energy flows:** The weight of the items, the quantity

manufactured, and the composition of the materials utilized are all given in this section.

- **Measuring operating and maintenance conditions:** Particularly in instances where preventative maintenance is directly tied to energy use, this section includes measures of plant availability times, as well as inspections of the proper operation and reliability of steam traps, measuring instrumentation, data logging equipment, burner nozzles, and engine lubrication, as well as visual and audible leak detection.

Two causes of mistakes have a direct effect on the precision and anticipatory capabilities of an energy standard:

- **Error in estimating the value of a meter or measurement:** This is any uncertainty regarding the quantitative or qualitative data used to construct the standard results in inaccurate energy savings forecasts.
- **Errors linked with the energy standard's structure:** Examples are an incorrect choice of a mathematical function or the omission of critical parameters from the standard's formula. Typically, the chosen function embodies some physical law but does not take all necessary decisive aspects into consideration.

The energy audit's objective is to minimize measurement/parameter estimation mistakes, as well as inaccuracies caused by inaccurate standard formulation. In terms of significance, the primary source of inaccuracy is typically due to inaccurate estimation/measuring of energy and mass flows, as well as a lack of measurements or information on the state of critical parameters.

1.4.2 METHODOLOGY FOR ESTIMATING ENERGY PARAMETERS

Typically, energy and/or production parameters are estimated using measurement methods. A suitable measurement method is chosen for each parameter under estimation, which may include one or more measurements of a single or multiple physical quantities. For example, the bare minimum requirements for estimating the energy flow of a boiler's exhaust gases are their temperature, their concentrations of O_2, CO_2, CO, and water vapor, as well as their mass flow.

Auditors may employ mass and energy balances to simplify their task without jeopardizing its correctness. Instead of monitoring exhaust flow, for example, fuel consumption could be measured, allowing for an indirect estimation of the flow via fuel and air mass balance. In other circumstances, such as when estimating heat losses from surfaces, direct measurement is not always practicable. In such instances, it is customary to measure another quantity (in this case, surface temperature), while the desired parameter (heat loss) is computed indirectly using a standard.

At least three times, these measurements should be made to allow for any thermodynamic imbalance or measurement mistake. A parameter's value should be replicated for all potential typical operating conditions under which the installation is expected to operate. As a result, in addition to conducting measurements, the Auditor should describe the critical parameters affecting the efficiency and specific energy use. The process of measuring and estimating, as well as the specification and calibration of instrumentation, should be based on relevant national standards, if they exist, or on international standards (e.g., CEN, ISO); otherwise, the Auditor should state explicitly the physical law or international standard upon which the estimation was based.

For as long as necessary, the Auditor may employ known and widely used nomograms, computational methodologies, and PC source codes (e.g., use of codes for the estimation of energy losses through walls or exhausts, based on temperature measurements). The Auditor's methodologies and tools, along with their sources and verified accuracies, should be explicitly stated.

1.5 PORTABLE MEASURING INSTRUMENTS

It is vital to employ reliable and complete data for a lengthy time period in order to measure and estimate the relevant parameters. In practice, however, complete data are rarely available. Additionally, existing measuring instruments are not properly maintained or calibrated, resulting in a lack of confidence. The Auditor should analyze the instruments' operating and maintenance state, as well as their potential for measurement error.

The Auditor arranges the measuring operations based on the audit's criteria and standards, utilizing both mounted and portable instruments. Measuring activities are typically conducted during the auditing phase due to their brief duration. As a result, measurements are made on a moment-by-moment basis rather than on a seasonal or annual one. In actuality, measurements are made in terms of power, not energy. Power is defined as energy per unit of time and is thus a "real-time" number that can be measured in seconds or minutes.

The Auditor should ensure that the measured system is in thermodynamic balance during the measurement activity, as shown by the stability of the instrument readings. As a result, measurements made using portable devices are ineffective at precisely estimating monthly and annual energy consumption patterns, as time is not directly monitored. Nonetheless, the usage of portable measuring equipment throughout the auditing process is highly beneficial for determining the efficiency of energy consumption installations, defining an energy consumption standard, verifying existing/installed instrumentation, and getting fresh data.

During an auditing session, the most often measured quantities are as follows:

- Flows of liquid and gaseous fuels
- Voltage, current intensity, and power measures, as well as the power factor
- Surface temperatures of solids and liquids
- Fluid pressure in tubes, furnaces, and vessels (including measurements under evacuation conditions)
- Exhaust gases as measured for their CO_2, CO, O_2, and smoke emissions and contents
- Relative humidity
- Luminance levels

The Auditor must describe from the start a list of all instruments necessary to complete the auditing process, whether available or not and/or portable or not.

1.6 MEASURING AND VERIFICATION (M&V) PROGRAM TO DATA SCIENCE

When it becomes clear that the intended precision cannot be achieved during the auditing phase due to time, measurement, technological, or other constraints not anticipated during the planning process, the Auditor develops an analytical program of measuring and verification (M&V). An M&V program is composed of an analytical description of the necessary instrumentation, measurements, typical operating conditions, and the methodology for analyzing the proposed data.

The M&V program's objective is to authenticate the estimations produced during the auditing method and to substantiate, at an acceptable level, the measurements and assumptions established during the auditing, for both reference consumption and the magnitude of prospective energy saving measures. Additionally, it tries to provide an objective system for monitoring and controlling the performance of the installation or its specific energy consumption. As long as requested, the M&V program may involve the installation of timers to monitor the operation of the installation. In practice, the Auditor should recommend a monitoring and verification scheme for each energy saving item indicated.

The duration of such an M&V program should be sufficient to allow an accurate depiction of average energy consumption before and after the energy saving measures are implemented, and this duration is determined by the nature of the required energy saving interventions. The annual reflected cost of an M&V program should not exceed 20% of the economic savings attributable to suitable energy conservation measures. It is often between 5 and 10%.

Additionally, the energy monitoring (EM) procedure necessitates the continuous or regular monitoring of a building, a complex, or a plant unit's energy behavior prior to and, in most cases, immediately following the implementation of energy saving measures on the shell or energy producing installations. As a result, it provides a technique of determining the efficacy of energy conservation measures by comparing the energy behavior of the building/plant before and during their implementation.

The following energy monitoring systems differ in terms of the depth of their energy audit coverage:

1. System that encompasses the entire building (single measurement area) and is based on current multimetering sensors, as well as electricity and fuel bills (a single energy measuring center)
2. System that encompasses the entire building (one measurement area) based on numerous partial measurements (many energy measuring centers)
3. Based on current multimetering devices, electricity and fuel bills, a system that covers independently some or all of a building's energy installations (multiple measurement areas) (a single energy measuring center)

4. A system that covers discretely some or all of a building's energy installations (multiple measuring locations)
5. Partial measurements (at multiple energy measuring centers) taken for each energy system auditing region

Energy monitoring systems are grouped into the following categories based on the degree of automation of the energy measurements:

1.7 MANUAL SYSTEM

A manual system includes that:

- A qualified operator monitors the counters' display.
- Data gathered from bills and invoices.
- Data tabulation is printed in calculated form.
- A theoretical evaluation is made of the specific area's energy behavior.
- A manually written report is generated.

SEMIAUTOMATIC SYSTEM

In a semiautomatic system, a certified operator monitors the counters' display and enters the data into a PC database. The data are processed automatically, and the auditing area's energy behavior is determined using a suitable algorithm. The auditing area's energy behavior is determined using a computer.

AUTOMATIC SYSTEM

In an automated system, instruments for measuring are connected to a personal computer. Data entry and processing are automated through the use of appropriate equipment and software. Advanced intelligent electronic systems are applied and interconnected with other information systems (in the case of a central building energy management system [BEMS]).

1.8 TYPICAL MEASUREMENTS AND INSTRUMENTATION

The following section summarizes the instruments that are most frequently used in M&V programs, whether portable or fixed, for measurements. The ones that provide an electrical output signal are more intriguing since they enable a PC to monitor measurements and gather data.

1.8.1 MEASURING ELECTRICAL PARAMETERS

The following instruments are included for measuring electrical parameters:

- **Ammeter:** Used to determine the amount of current absorbed by appliances and motors

- **Voltmeter:** Used to determine the voltage or voltage drop across a grid or electrical circuit
- **Wattmeter:** Used to determine the instantaneous power consumption of appliances/motors or the power performance of generators
- **Cos (φ) meter:** Determines the power factor of a circuit or checks the operation of rectification devices
- **Multimeter:** Used to determine all of the preceding values

All of these instruments are typically portable. They are connected to the wire by nippers and may include a data logger. Electrical power and energy consumption should be monitored in all places and installations that consume a lot of energy. Given the low cost of these sensors, it is prudent to consider their permanent installation in some of the aforementioned instances.

When all of these values are measured, a clear distinction must be established between total power (metered in KVA) and active power (often metered in kW), as well as between cos φ. Additionally, caution is required when electrical loads are not expected to exhibit a sinusoidal waveform, as is the case with variable speed motors and uninterruptible power supplies. The typical measuring instrument is based on a sinusoidal waveform, which produces inaccurate measurements. In these instances, the usage of meters capable of measuring true RMS (root mean square) values is required. Because the function of these meters is based on digital sampling, they can be replaced with PC-based meters.

Electrical properties can be measured with the aid of a sophisticated device called a power analyzer. After properly connecting the instrument to the electrical panel of the machinery or substation under evaluation, the instrument's display shows measurement readings for each phase and for the overall voltage, current, apparent reactive and active power, cos φ, and energy consumption. Every 20 seconds, the instantaneous measurements are repeated. Additionally, measurements can be stored in a memory pack for a lengthy time interval.

The technique for connecting a power analyzer to an electrical panel is often specified in the instrument's manual. It is critical that the instrument is connected properly (in this case, the display indicates "OK") in order for the measurements to be accurate. If possible, a temporary power outage should be performed, and the electrician in charge, in collaboration with the auditing staff, should complete all instrument connections (voltage connectors and ammeter clamps).

The memory pack, data are processed and analyzed using a software package given in the instrument's manual. Graphs presenting the power demand and cos φ fluctuations for the auditing period are automatically generated from the processed data. Additionally, power consumption of the item under test is recorded in kWh during the same period, as is reactive load per phase and for all three phases.

1.8.2 Temperature Measurement

Temperature meters with a PC interface are already available at the relevant retailers. The most frequently used measurement technology are as follows:

- **Detectors of resistance thermometers (RTD):** From the most cutting-edge technological tools, these include internal calibration and reset signals.

They are extremely accurate and are frequently used in M&A applications as permanent equipment.

- **Thermocouples:** Frequently used and reasonably priced, thermocouples typically cover a wide temperature range, from a few degrees to 1000°C, and are portable. They must be calibrated on a regular basis using specialist devices. Their primary disadvantage is a weak signal that is easily influenced by industrial noise.
- **Thermistors:** These are inexpensive permanent meters. They exhibit a robust, linear variation in response to the temperature signal and are capable of self-resetting. Nonetheless, thermistors and thermocouples are uncommon in M&V setups.
- **Infrared thermometers:** These determine temperatures at a distance by measuring the thermal radiation emitted by substances. They detect "hot spots" and areas of poor insulation. They are portable and simple to use, but their precision is restricted; they also require knowledge of the emissivity coefficient.
- **Traditional bulb thermometers (e.g., mercury types):** These are accurate enough to be employed in situations requiring isolated measurements.

It is worthwhile to describe the **infrared camera** and associated measurement techniques in further depth, as they are critical to the energy auditing approach. Thermal imaging or thermal inspection is a technique for determining the forms of thermal losses in structures, for example:

- Loss of the building's shell.
- Losses from ventilation holes.
- Losses in warehouses and fluid networks (for water, air, steam, etc.).

Infrared thermometry's fundamental premise is founded on the fact that everybody emits infrared radiation, which is entirely dependent on the body's temperature and the emissivity of its surface. The infrared camera is equipped with a sensor that transforms, via appropriate software, infrared radiation to voltage difference and, accordingly, to a picture with a color spectrum corresponding to the radiation levels.

The infrared camera captures the radiation emitted or reflected by a body's surface and converts it to visible colors for display. Each color represents a specific temperature range, allowing for direct temperature readings. The camera's temperature measurement is dependent on several camera parameters but also on the emissivity of the body's surface and color.

Prior to auditing heating systems (i.e., boilers), the systems should be operating at normal temperatures to ensure that the data from the measurements is as representative as feasible. The camera should run for approximately 5 minutes prior to any measurements being made to allow for automated self-calibration. To evaluate heat losses through the building's shell and the spots where insulation is deteriorating, the indoor temperature should be sufficiently higher than the exterior temperature. Thus a chilly and foggy day should be chosen to avoid incident radiation heating the walls.

Additionally, the following methods are advised for conducting temperature measurements on-site:

- **Calibration:** Thermometers must be calibrated prior to use and on a periodic basis thereafter.
- **Surface temperature measurement:** The sensor should be shielded from air conductivity with an insulating cover and kept away from radiation sources such as the sun, various radiators, or windows. Additionally, sensors should not be placed near air inlets or outlets. If a typical or average temperature is required, the sensor must be placed away from nonrepresentative locations, such as thermal bridges. As a substitute, numerous measurements might be taken in multiple representative places and the average of these results determined.

Infrared sensors or cameras can also be used to determine the surface temperature. As with surface temperature measurements, the sensor must be shielded from radiation sources such as radiators and windows. Measurements in a space must be taken at least three times at three different heights (stratification) and in the space's center, to prevent as much as possible the influences of surrounding surfaces.

1.8.3 FLOW MEASUREMENTS

The infrared camera captures the radiation emitted or reflected by a body's surface and converts it to visible colors for display. Each color represents a specific temperature range, allowing for direct temperature readings. The camera's temperature measurement is dependent on several camera parameters but also on the emissivity of the body's surface and color.

Prior to auditing heating systems (i.e., boilers), the systems should be operating at normal temperatures to ensure that the data from the measurements are as representative as feasible. The camera should run for approximately 5 minutes prior to any measurements being made to allow for automated self-calibration. To evaluate heat losses through the building's shell and the spots where insulation is deteriorating, the indoor temperature should be sufficiently higher than the exterior temperature. Thus a chilly and foggy day should be chosen to avoid incident radiation heating the walls.

1.8.4 MEASURING AIR HUMIDITY

Typically, air humidity is determined using dry and wet bulb thermometers. These measurements take time and demand caution throughout the preparation phase. Electronic meters with rapid response have been created recently but are limited to temperature ranges of up to 60°C. To be more precise, the following instruments are used:

- **Psychrometer (or wet and dry bulb thermometer):** This is made up of two temperature sensors, one of which is coated in a cotton cloth moistened with distilled water and which records a temperature close to the thermodynamic temperature of a wet bulb. The relative humidity can be calculated using the dry and wet bulb temperatures and the barometric pressure. Psychrometers are inoperable when the ambient air temperature is less than 0°C. They require frequent cleaning and replacement of cotton cloths. If the relative humidity is greater than 20%, accuracy is approximately 0.5 K when properly maintained. A lithium chloride cell can be used in place of the psychrometer. It is a straightforward and reasonably priced instrument with a function range of −29 to +70°C and a precision of 2 K. Air velocity greater than 10 m/s can cause the calibration scale to shift, and exposure to high humidity combined with a power loss (for example, a power breakdown) can dissolve the salts, necessitating instrument rehabilitation.

- **Ion exchange resin sensor (pipe type):** Due to its rapid reaction and capability for continuous measurement, this reasonably priced type of sensor is frequently employed in humidity meters for the detection of relatively constant airflow. The pipe-type sensor operates at temperatures below 75°C and is extremely sensitive to organic solvents (e.g., oil vapor) and polystyrene glues. Certain sensors include a metallic filter to protect them from the majority of airborne particles. A few minutes of exposure to excessive humidity can result in a loss of calibration or possibly sensor failure.
- **Digital humidity meter:** This portable device measures both ambient humidity and the moisture content of a wide variety of structural materials, such as bricks, wood, coatings, and sand. More precisely, the instrument indicates the degree or level of humidity present in the substance. When the air relative humidity is high or the surface under measurement appears to be moist, materials or surfaces can impair the measurement's accuracy. It is composed of two components: the primary device and the sensor. Its operation is straightforward and comparable to that of electronic thermometers.
- **Thermohydrographs:** These are used to simultaneously measure and record both air temperature and relative humidity. The temperature sensor is composed of a highly processed bimetallic strip designed for long-term operation, the expansions and contractions of which are communicated to the pen's arm via a unique mechanism. The humidity sensor is made out of a bundle of human hair, and any change in its length caused by changes in humidity is transferred to a graph via the appropriate pen. A translucent cover enables continuous surveillance. The temperature range is −15−+65°C, and the humidity range is 0–100%.

1.8.5 Exhaust Gases Measurements

Exhaust gases measurements are necessary for determining the efficiency of combustion in boilers, furnaces, and burners. They include measurements of CO_2, CO, SO_x, NO_x, smoke content, and temperature. Historically, these measurements have been made using inexpensive portable instruments. Today, electronic gas analyzers are available that can instantly test all of these values while also doing combustion efficiency estimates. These meters are calibrated and reset automatically.

Comparisons of these electronic instruments to conventional instruments must be made with caution. The latter measure the dry gas circumstances, whereas the electronic versions monitor continuously and in real time. Additionally, prior to doing any on-site tests, the boiler should be operated for a period of time to reach its normal operating temperature.

The gas analyzer's sample probe is put into the chimney, and its end must be located in the center of the exhaust gases' core (the middle of the chimney). This is possible with precision because current gas analyzers are capable of displaying the temperature of each spot at which the probe is positioned. As a result, the optimal sample point is placed directly above the point of highest temperature, in the center of the gas flow.

After proper sampling, the gas analyzer analyses the gases and determines the percentages of CO, CO_2, O_2, SO_2, NO_x, and C_xH_x in the exhaust gases using built-in algorithms. Modern gas analyzers are completely automated, and when the right sample is taken, the boiler efficiency and percentage of the target gases are displayed on the gas analyzer's display. A gas analyzer can collect instantaneous measurements, but it can also provide an average over the duration of its connection to the boiler.

Additionally, the smoke index can be determined using the same device. By inserting a particular filter paper into the gas analyzer's probe, the value is presented directly as a smoke index. This is accomplished visually by placing the special filter paper beneath the smoke index's comparing scale (Bacharach scale), such that the smoke spot completely covers a hole in the comparing scale. The smoke index is calculated by comparing the closest blackening level region to the smoke spot on the comparison scale.

1.8.6 MEASURING OPERATION TIME

In many circumstances, it is required to continuously monitor the operation hours of an appliance or installation, as well as the time period during which it is operating. The second instance necessitates the usage of a data logger. This measurement is required in order to determine the amount of energy saved. As a result, these meters are among the first that the Auditor recommends for installation as a critical measure of instrumentation update.

Other Necessary Measurements

Additional measurements that are typically made during the auditing method include the following:

- **Luminance level measurements** are used to identify lighting regulation violations. Luminance-measuring equipment employs a light sensor equipped with a unit for correcting for color and incidence angle. The sensor should be connected to an analogue or digital display through a flexible connection for optimal performance. This minimizes the possibility of shading the sensor during measurements. Measurements should be conducted under controlled conditions (a time period should be given for the lamps to preheat). It must be established that natural illumination has no effect on the electrical lighting measurement.

- **Total dissolved solids (TDS)** measurements in the boiler's water allow for monitoring the water treatment system and optimization of blowdown water volumes.
- **Pressure measurements (static or total) on fluids** are used to verify an apparatus's operational condition (e.g., boiler exhaust pressure measuring or temperature differences adequate to install heat exchangers for heat recovery). The measurements are used to monitor the condition of steam traps in order to locate and repair or replace damaged units.

Finally, measurements are required for the accurate determination of **energy losses** that are not typically performed during an energy auditing procedure, such as the measurement for determining the overall thermal loss coefficient, which is performed using a unit consisting of the main device, space and surface temperature sensors, and a heat flow sensor for the detection (q). A resistance is provided on a 16-pole terminal to which the sensors' terminals are attached in order to compensate for a reference temperature in parallel.

Indoor and outdoor temperatures are measured via instrument sensors. If a discrepancy between these two parameters is detected during measurements, the equipment determines the temperature difference affecting the heat flow. Additionally, a plate is utilized to obtain data for the structural element's surface heat flow density ($q = Q/F$). The total time required to conduct readings on any structural element is 120 hours. Throughout this time period, mean temperature and heat flow density readings are kept every 2 minutes, and a mean is calculated and stored separately every 2 hours. Each new value of the heat loss coefficient k is calculated as the mean of all previous values.

In the measuring of ventilation loss, the ventilation air quantity in a room is difficult to calculate analytically because it is dependent on the permeability of the building shell, the arrangement of the partition walls, the difference between the temperatures of the indoor and outdoor spaces, the direction and velocity of the wind, the construction quality and design, and other factors. Two methods are typically used to determine the ventilation of buildings:

- **Tracer gas approach:** This has no effect on the interior or outdoor circumstances within the space under measurement. This method can be used to determine the dimensions of small or medium-sized structures having a clearly defined shell. The tracer gas must be evenly distributed throughout the room, the barometric pressure must be consistent, the tracer gas inflow must not cause fluctuations in air density, and the interior air temperature and density must be constant.
- **Pressure method:** This is used to determine the airtightness of a structure's shell. It is a simple and reliable short-term procedure that is achieved by mechanically over- or underpressurizing the interior of the audited space. A fan is necessary to create a pressure differential between the room and the surrounding environment (blow-door), a manometer is required to monitor

the pressure difference, and a flowmeter is required to measure the airflow created by the pressure difference from the building's shell.

Finally, in the case of energy projects such as RES, a particular measuring setup is required in order to precisely estimate energy production and system performance. The instrumentation required varies according to the nature of the RES and the technology used. What is often measured is the energy source's intensity as well as the pertinent meteorological parameters that affect the RE source in question.

1.9 REPORT WRITING

After the detailed auditing, the Auditor will give an energy audit report covering the potential areas of energy conservation and the Auditor shall work hand in hand with the company engineers. The energy audit report will be divided into:

- The consolidated list of energy saving opportunities.
- Details of energy saving opportunities.
- Organizational aspects of energy conservations and implementation program.
- Formats for monitoring energy saving opportunities.
- Contains an annex giving details of back calculation.

1.10 ELECTRICAL SYSTEM LOSSES

The losses in electrical energy systems can be broadly classified as follows:

- Electrical losses in power generating, transmitting, and distributing equipment
- Mechanical losses in rotating systems
- Corona losses in EHV voltage levels and higher
- Transmission and distribution line/cable losses
- Losses due to failure/maloperation in operation of protective relaying systems and switchgear
- Losses in cooling systems
- Metering losses
- Commercial losses
- VAR loading losses
- Loss due to interruption of supply and breakdown.
- Loss due to improper or/and inadequate maintenance
- Losses due to poor earthing
- Losses due to overloading
- Losses within consumers/industrial premises

1.10.1 ENERGY BALANCE SHEET

The balance sheet prepared by an Energy Auditor after conducting energy accounting studies is called the energy balance sheet. By investigating the energy balance sheet in detail, the sections of power network having more line loss values can be identified visually, and suitable steps for reduction according to the necessity can be taken.

Reduction of Technical Losses

When power is generated, transmitted, and distributed from one place to another, certain losses occur due to the inherent characteristics of the system. These losses cannot be totally eliminated, but it is possible to reduce them, with considerable investments. The Government of India requires the various electricity boards/power utilities in the country to take effective steps to bring down the T&D losses further to the level of 12%. A reduction in 1% of T&D losses results in a savings of 4638.9 lakh units of energy to the SEBs every year.

1.11 MECHANICAL LOSSES IN ROTATING SYSTEMS

- A quantum of the converted energy is lost mechanically in generators in the prime mover due to improper mechanical coupling, misalignment, etc.; in the case of drives, power output is reduced. Further, the misalignment leads to magnetic imbalance inside the electrical machine, thus straining the bearings and thrust pads, which in turn produces more heat in addition to the heat produced in the rotor.
- In bearings and thrust pads, due to routine wear and tear, faults, and peeling of white metal, the mechanical losses increase by improper labyrinth clearances.
- In rotors, losses occur due to improper static and dynamic balancing.
- Losses occur in bearings, thrust pads, mechanical couplings, flywheels, etc. due to vibration. Also, harmonics in the rotor flux and stator currents due to load conditions and faults may result in vibration.
- Windage loss results from fan loss, ventilation losses.
- Friction loss can be caused by improper lubrication, inadequate preventive maintenance, or any of the reasons previously listed.
- Inadequate lubrication oil levels and pressure in the lubricating oil system can lead to losses.

1.11.1 REDUCTION METHODS

- Proper mechanical coupling after making the correct alignment is one reduction method.
- After conducting a vibration analysis on the entire machine, static balancing at the manufacture's premises and dynamic balancing have to be done on-site at the rated speed during erection, as well as periodical dynamic balancing, and aged, worn-out bearings and thrust pads have to be replaced.
- Labyrinth (oil film) clearances have to be maintained, the condition of lubricating oil and oil strainers has to be inspected, and the condition of DE and NDE bearings, thrust pads, and thrust bearings has to be checked periodically for free movement of rotor, which in turn reduces rotational losses.

- Light and strong cooling fans have to be mounted for smaller and medium-sized hydrogenerators for effective cooling and reduction in windage losses. The atmospheric openings in ventilation ducts and in manhole/inspection covers have to be sealed tightly in the case of open- and closed-circuit ventilation systems. All the cooling fans should be switched on in the ventilation system to create the designed airflow in the ducts, and upon failure of any fan, replacement shall be made immediately.
- Maintaining adequate lubricate oil levels and pressure in the lubricating oil system for nondriven and driven end bearings for generators, thrust pads, turbine bearings.
- The uniformity of the air gap in every stage of erection at all positions of the rotor poles inside the stator needs to be checked.
- During erection and maintenance, proper care should be taken to keep up the dimensional tolerances within limits, tightening all bed bolts/nuts, and checking oil film clearances, white metal thicknesses in the case of bearings and thrust pads, rotor/exciter air gaps, correct size/capacity of motors, along with appropriate starters for all auxiliaries/drives, etc.

1.12 DIFFERENT TYPES OF LOSSES IN A FACTORY

1.12.1 Losses in Electrical Machines

The various types of losses and their reduction methods in electrical machines such as synchronous machines, induction machines, and transformers are discussed next.

Reduction Methods—Alternators

- In turbogenerators, where there is a long radial air gap, it is possible to eliminate stator slotting altogether and to fix the winding to the smooth stator bore while still leaving adequate rotor clearance. In this type of spotless winding, mechanical and thermal savings in the core and the insulating materials are made, along with a reduction in core losses.
- Reduction results from the proper selection of rotor core material with desirable B-H (flux density–magnetizing force) characteristics.
- Reducing VAR loading and controlling excitation current both help.
- Finally, it is also helpful to use modern static excitation equipment/brushless excitation systems by replacing conventional age-old excitation system consisting of PMG, pilot exciter, and main exciter.

Reduction Methods—Induction Machines

- Avoidance of operation at low loads and overloading effects
- Avoidance of operation with larger slip values
- Use of energy efficient designed induction machines
- Use of intelligent controllers
- Proper and correct selection of motors according to applications
- Use of soft starters for motors with intelligent controllers

Reduction Methods—Transformers

- The core loss in transformers can be reduced by the selection of a proper core material, such as CRGO steel core instead of HRGO steel. Also, the use of amorphous core instead of CRGO core will reduce the core loss.
- The dielectric loss in transformer oil can be reduced by filtering and reconditioning with an oil filtration plant.
- Spraying varnish on the insulation can minimize the dielectric loss in core losses of a distribution-type transformer used for low-voltage applications.
- To reduce iron loss as well as to improve all-day efficiency, the load curve has to be flattened by studying the load pattern and rescheduling loads consumption.
- The cooling fans and oil circulation pumps have to be maintained and made available to the power transformer under loaded conditions; the control circuits for switching "on" and "off" cooling fans and pumps have to be periodically checked for proper functioning.
- The usage of booster transformers or of tap changing transformers for voltage control in a supply network reduces losses and is also required for:
  - Adjustment of consumers' terminal voltage within the statutory limits.
  - Control of active and reactive power flow in the network.
  - Seasonal (5–10%), daily (3–5%), and short period (1–2%) adjustments in accordance with the corresponding variations of load. These types of booster transformers are used for control of reactive power flow in interconnecting feeders in distribution networks.
- Harmonic current compensation.

Reduction Methods—Corona Loss

- Use of bundled conductors (duplex, triplex, etc.,) reduces the loss due to an increase in effective conductor diameter.
- Loss can be lessened by designing and selecting proper towers according to geographical and climatic conditions with adequate spacings/clearances between conductors and ground clearances.
- Opting SF6 gas insulated substations (GIS) can reduce corona loss in substations with voltage levels above 230 kV.
- Corona loss can be reduced by using HVDC and FAC transmission for EHV voltages above 400 kV.

Reduction Methods—VAR Loading

- Provide capacitive compensation at all voltage levels.
- Provide step-less compensation through solid-state power factor compensators and controllers.

Reduction Methods—Transmission and Distribution (T&D) Line Losses

- The length of long lines has to be reduced at lower voltage levels, i.e., distribution level. Overloading on the existing lines caused due to load growth has to be avoided since the squared value of the current increases the loss. In other words, overloading by 10% increases the losses by 21%.

- The new substations have to be energized at the load center, and the existing ones have to be relocated closer to the load center.
- Strengthen the existing size conductors with higher-sized conductors.
- Bifurcating/trifurcate the existing overloaded feeders.
- I^2R losses in LT lines can be reduced by a reduction of the HT/LT ratio to 1:1, for example, by reducing the LT (415 V, three-phase) line lengths and increasing the HT line length. Additional distribution transformers can be installed in the load centers to achieve the desired HT/LT ratio in order to reduce LT line loss.
- Installing load manager and feeder manager software, intelligent control and load flattening can be achieved by redistributing loads.
- Another reduction method is to ensure proper and adequate earthing of neutral terminals and equipment after measuring the earth resistance value.
- Select HVDC transmission systems for medium-distance transmission lines, EHV AC transmission systems for short lines, and FAC transmission systems for long lines.
- To reduce the dielectric losses in cables, bus ducts can be used for indoor applications. Use of XLPE and OF cables leads to reductions in dielectric losses.
- By conducting tan delta tests on EHV cables periodically, the polarization index value can be measured, which in turn reflects the quality of the insulation. If the value exceeds specified limits, the cables need to be changed, thus ensuring protection as well as reducing the dielectric losses.

1.12.2 METERING LOSSES

Metering loss is present at all voltage levels. The various types of metering losses encountered in a typical electrical system are:

- Losses in conventional electromagnetic meters due to low recording of energy consumption under lightly loaded conditions, which occurs due to the inability of the current coil to move the recording mechanism.
- Improper selection of meters, current transformers, and potential transformers due to urgency and nonavailability.
- Inaccuracy in instrument transformers due to incorrect selection of accuracy class.
- Ratio error due to core saturation of current transformers.
- Human-made errors like inaccurate/wrong calibration, adopting incorrect multiplication factors for CTs and PTs, reversing the phases, or changing the polarities and loads, exceeding the burden of instrument transformers.

Reduction Methods

- Using high-quality meters with current compensation for overloads and voltage compensation for low loads
- Recalibration and periodical testing of meters with full load and low load and at lagging/leading power factor loads

- Reducing the burden on the instrument transformers by using shorter control cables with higher cross-sectional copper conducting area and replacing all electromagnetic relays and meters with static or numerical ones
- Use of static TOD programmable energy meters with sequential events recording facilities along with provisions to read all electrical parameters involved (The parameters and events recorded in the memory of the microcontroller or microcomputer of the meter can be used for analyzing and rectifying meter errors.)
- Meters measuring of THD (total harmonic distortion) for inductive-type loads

1.12.3 Losses due to Protection Failure/Delay/Maloperation

Delay in the operation of the protection system due to electromagnetic relays—delay in clearing the fault—is crucial factor because the switchgear, power transformer, and other equipment are subjected to severe strain.

Reduction Methods
- Use of numerical and programmable relays
- Adequate earthing
- Main and backup relays
- Voltage/current/time grading of relays
- Use of high-speed tripping breakers
- Periodic maintenance and checking/testing of relays, protective system, and switchgear for their intactness

Commercial Losses
- Loss of revenue from unmetered free services such as agricultural and HUT services
- Loss of revenue due to theft of energy indulged in by electricity users
- Misuse of electricity tariff
- Unauthorized load
- Violating Indian Electricity Act and rules
- Incorrect or misappropriate application of tariff
- Improper or delay in sending electricity (energy consumption charges) bills by the power distributing authority

Reduction Methods
- Metering of agricultural and HUT services
- Watching the consumption pattern of electricity users periodically and taking suitable action on drops in consumption
- Identifying theft-prone areas and conducting surprise inspections
- Conducting mass raids
- Forming special squads for surprise inspection of consumers' premises
- Conducting periodical and routine inspections
- Periodical review and analysis of electricity bills and revenue collection using intelligent software

- Hourly updated centralized billing and collection software
- Using prepaid electricity meters

1.12.4 Losses in Heat Dissipation Systems (Cooling Systems)

- Reduction of output in power transformers due to failure of cooling fans
- Replacement of failed motors by higher- or lower-capacity motors in cooling fans or pumps, resulting in over- or undercooling effects
- Reduction of output of generators due to failure of air cooling fan inside the machine, due to reduction in flow rate of water coolers to stator by blockages in water pipeline (formation of scales from hard water, corrosion of pipes, etc.,), reduction in waterhead, etc.

Reduction Methods

- Immediate replacement of failed cooling fans or pumps
- Use of soft water by installing water softening plants with closed-circuit cooling towers
- Painting the cooling water and other coolant pipelines with anticorrosive paints
- Using anticorrosive solvents along with the cooling water
- Maintaining adequate coolant pressure head
- Avoiding overloading and reactive power loading to reduce temperature rise.
- Using amorphous core and dry-type transformers

1.12.5 Losses due to Supply Interruption and Switching

- Losses due to forced and planned outages
- Losses due to low reliability
- Losses caused by increased switching operations

Reduction Methods

- Doing adequate preventive, periodical maintenance without any lapse can minimize forced outages. Also, by adopting scientific preventive maintenance schemes, installing lower-maintenance, standard, BIS or ISO mark equipment, planned outages can be reduced.
- By proper planning, scientific operation and control, system reliability can be improved.
- The switching and fault-clearing operations can be reduced by distribution automation.

1.12.6 Losses due to Overloading

- Losses result from squared I^2R losses, low power factor, and poor voltage regulation.
- On overloading, the rate of failure of the equipment is increased.

- Overloading also results in increased noise, more vibration, and hum in electrical equipment.
- Reduction in the life span of equipment leads to losses.

Reduction Methods

Avoid overloading by load modeling, planning, managing, and meeting the demand by installing new generating stations.

1.12.7 Reducing Losses by Proper/Adequate Maintenance and Earthing

- Carrying out proper periodical preventive maintenance
- Replacing failed and weakened lightning arrestors
- Ensuring perfect neutral and body earthing of all equipment
- Carrying out regular cleaning, lubrication, painting, checking alignment, and other work

1.12.8 Reducing Losses by Proper Grid Management

- Installing under-frequency relays in all EHV feeders and in all interstate feeders at the 110 kV level and above to reduce overloading of feeders and preventing the tripping
- Using ABT (availability-based tariff) meters for the feeders to watch and control the load flow parameters and to penalize the utility that draws power deviating from the norms fixed by the grid management body or Regional Electricity Board coordination committee
- Installing so-called cold-start generating stations for energizing the grid during blackouts and having a greater number of such stations in vital load centers
- By constructing pumped storage hydroelectric schemes to match the generation/load demand curve
- After conducting power flow analyses on the entire grid at all buses in generating stations and at all transmitting and distributing substations, conducting load and energy forecasting and planning to meet load growth and to reduce overloading
- Remote-controlled generating stations and substations
- Real-time monitoring and control of substations
- Resorting to load management by installing intelligent load managers in feeders
- After conducting short circuit studies, installation of an additional or second backup protection system in fault-prone areas and where equipment frequently fail

1.12.9 Losses in Electrical Equipment in Consumers' or Industrial Premises

- Losses in lighting loads due to incorrect positioning of fittings, using of conventional chokes, luminaires without reflectors, etc.

- Losses in wiring circuits due to common power and lighting wiring, air conditioners, motor starters and conventional-type and static-type variable-speed drives
- Improper selection of pumps and motors for the specific application
- Replacing under- or oversized motors during failure
- Improper and poor load management
- Thermal losses caused due to no or poor insulated refrigerant or steam or hot water pipes, which in turn increase the electrical energy consumption

Reduction Methods

- Using lux meters and selecting the suitable type of light fittings or luminaires with reflectors
- Using electronic ballasts and compact fluorescent lamps (CFL)
- Painting the walls and ceiling with a suitable color that reflects maximum light
- Using soft starters with intelligent controllers for motors
- Ensuring perfect contacts between copper and aluminum (thus minimizing losses and heating due to bimetallic action)
- Insulating refrigerant and hot water pipes of air conditioners and heaters/geysers, respectively, and varying the temperature according to the ambient temperature with a temperature controller
- Selecting a suitable pump for an application as per the pump chart
- Use of energy efficiently designed motors with longer shafts and smaller diameters
- Not using under- or oversized motors
- Using a programmable, single, intelligent controller for controlling all the drives in a machine according to the processes/materials/products in the pipelines
- Flattening the load curve with proper load modeling and planning

Modern Techniques Used for Reducing Losses

Nowadays, a number of scientific, modernized, automated techniques are applied throughout the world for optimization of energy savings, investments, and the payback period, and some are discussed next.

Distribution Automation

As distribution losses represent the major segment of electrical energy losses (7%), the automation of power distribution sector will lead to drastic cuts in total losses. The techniques used are:

- Installing photoelectric or light-dependent resistors (LDR) for outdoor and street lighting controls.
- Installing line post sensors in LT and HT lines for detection of faults and signaling to fault detectors.
- Installing online real-time fault detectors and analyzers.
- Installing load and feeder managers in all HT feeders at substations.

- Installing load flow controllers and load pattern analyzers with remote-controlled molded current circuit breakers (MCCB) or static transfer switches (STS).
- Installing remote telemetry units (RTU) for monitoring load flow in distribution substations.
- Remote controlled line and capacitor switching controls for switching off and switching on LT lines and capacitors based on load flow and reactive power flow.
- Power line carrier communication (PLCC)–operated substations and feeder controllers.
- Microwave/satellite communication-controlled substations and feeder controllers.
- Studying all the electrical parameters of electrical power flow at a glance by installing power analyzers at all distribution substations and downloading the data recorded in the analyzer for a specified period (This is very useful in load modeling, load redistribution, and load windowing for flattening the load curve.)
- Reducing restoration timing on fuse-off calls and supply interruption timing by installing an online consumer information service system
- Erecting a ring mains distribution network and having alternate and standby supply points, restoring electrical supply immediately upon faults
- With the help of a geographic information system—a database of the LT and HT lines network of the distribution system mapped onto a local geographic district or taluk map—interfaced to remote switching stations, making fault location and diagnosis easier (The faulty section is instantly detected, and the supply is restored to all other areas except the faulty section. With a ring main network, the supply to the areas after the faulty section is also resumed immediately.)
- Installing remote ripple or/and harmonic controllers for electrical services (for example, induction arc furnaces) prone to introduce ripples or harmonics in the electrical network and to trip the sealed remote controlled circuit breakers (RCCB) on such introduction (This ensures that power quality is protected.)
- Introducing multirate meters for measuring power at different times of the day with different tariffs—say, three tariffs, namely peak, normal, and off-peak—and encouraging consumers to use the off-peak power by fixing a lower tariff for it, thus flattening the load curve and bridging the load generation gap.
- Sealing the aerial fuses in the line post and meter fuses in the meter board, thus preventing the connecting of unauthorized additional load since these fuses have to be replaced by the staff of the power utility based on the complaint/entry made by the consumer in the online consumer information service system
- Erecting self-protected-type distribution transformers and intelligent universal transformers

Other Techniques
- Replacement of insulation in aged and end-of-life-span machines
- Replacement of conventional electromagnetic relays in protection systems by numerical relays with intelligent programming facilities

- Using computer-aided software such as:
  - SCADA—Supervisory Control and Data Acquisition System
  - EMS—Energy Management Software
  - CAPSI—Computer Aided Power System Improvement

1.12.10 Air Conditioning System

The equipment under air conditioning system covers the chiller compressor:

- Auxiliary consuming equipment—pumps and fans.
- Air handling units.
- Distribution and utilization system.

The scope of the study includes:

- An estimate of actual TR (tons of refrigeration) generation by the chiller based on BIS standards.
- Measurement of power consumption and estimation of specific power, i.e., kW/TR for A/C compressors, chilled water [umps, condenser fans, AHU fans, etc.
- Operational features of the compressor, such as temperature, pressure control.

A study of the auxiliary component for energy efficiency and its impact on the auxiliary component includes:

- Chilled water pumps (measurement of water flow is possible, power, condenser fans (measurement of flow, power, AHU fans)
- Study of distribution network for temperature drop, pressure drop, insulation
- Detailed study on AHU's airflow rate measurement across the AHUs, measuring the temperature and relative humidity in AHUs (supply, return and fresh air); estimation of TR generation by AHUs, speed measurement of AHUs fans—if applicable
- Measurement and recording of temperature and relative humidity in critical areas by using microprocessor-based tiny tags
- Monitoring of operating parameters vis-à-vis desirable parameters in user locations
- User area analysis to arrive at energy conservation measures
- Application potential for various energy saving retrofits

1.12.11 Pumping System

If the facility has water pumps for chilled water circulation, raw water pumping, and various applications, the study of pumps (7.5 kW and above) would cover water pumps:

- Flow and head measurement of water pumps by using sophisticated energy audit equipment
- Measurement of power parameters (kW, KVA, PF, frequency, current, voltage)

- Evaluation of efficiency of pumps
- Measurement of speed by stroboscope
- Application and matching of drive
- Application of flow control methods
- Application of retrofit for energy savings

1.12.12 FANS AND BLOWERS

The study of fans and blowers would cover the system having 7.5 kW and above and include the following:

- Flow and head measurement by using sophisticated energy audit equipment
- Measurement of power parameters (kW, KVA, pf, frequency, current, voltage)
- Evaluation of efficiency of fans and blowers
- Measurement of speed by stroboscope
- Application and matching of drive
- Application of flow control methods
- Application of retrofit for energy saving

1.13 ENERGY AUDITOR

An Energy Auditor is one who is authorized by the government to carry out the energy auditing exercise and to suggest, implement, and supervise the energy conservation program in the electrical installations of a consumer. It is a mandatory that all power consuming services have to be energy audited. The duties of an Energy Auditor are to:

- Provide a computerized energy audit that is acceptable to the Department of Energy.
- Standardize the audit procedures.
- Physically examine the field.
- Visually draw plans for the industries/dwellings that are being audited.
- Place objects such as doors, windows, other openings, etc. and specify their attributes for calculations.
- Identify measures for specific objects that are included on the drawings.
- Generate a prioritized savings-to-investment ratio that ranks the cost-effectiveness of measures

1.13.1 PURPOSE OF ENERGY AUDIT

The purpose of an energy audit program is to systematically identify opportunities for saving energy in the plant and to realize cost reduction benefits. It also involves identifying the loss of energy, retrofitting, and modification, alternate energy source, etc. Our experience shows that 5–25% energy can be saved for any plant by carrying out systematic and scientific energy auditing. The establishment of a realistic goal for overall long-term energy saving should be carried out to assure that the proposed energy conservation programs conform to the functions and primary operation of the plant.

1.14 SPECIFIC ENERGY CONSERVATION OPPORTUNITIES

1.14.1 ENERGY CONSERVATION IN COMPRESSED AIR SYSTEM

A leak test in the compressed air system should be done in the detailed energy audit. To assess the amount of energy that can be saved in compressors, it is essential to know their operating efficiency. This can be found by performance testing compressors as per the British or Indian standard. Energy conservation measures can be suggested for compressors only after evaluating their operating efficiency. This is done in the final audit.

1.14.2 ENERGY CONSERVATION BY OPTIMIZATION OF DM PLANT

A demineralization (DM) plant with a mixed bed generates polished water having a silica level of 0.02 ppm. The maintenance of the silica level is very important criterion for the efficient running of turbines and for reducing the chemical used for the regeneration of resins. Studies on the regeneration of resins, yield, and specific consumption of a chemical is a must and are included in the detailed energy audit report.

1.14.3 OPTIMIZATION OF BOILER/STEAM SYSTEM

For a waste heat boiler generating steam of 7 T/hr at pressure of 22 kg/cm, for example, there is no steam flowmeter to monitor the quantity of steam. Steam flowmeters have to be installed at the generation and consumer ends. Steam balance is required in steam generation. The design is included in the final audit after conducting measurements and testing.

1.14.4 ENERGY CONSERVATION IN FANS/BLOWERS

There is good scope for energy conservation with fans. Performance testing of fans and blowers is to be done as per the BS standard. In flow measurement, reliable values of tapping at the appropriate place at the suction and discharge of the fans are required. The tapping for a rectangular duct should be as per the log-Tchebycheff rule and for the circular duct as per the log-linear rule. The calculation for the airflow rate, air HP, and the efficiency should follow the British standard for testing fans taking, into account various corrections factors. Only then can accuracy be ensured. The readings taken with a few measurements lead to erroneous results since the flow pattern varies all along cross section of the duct.

1.14.5 ENERGY CONSERVATION BY CONVERSION OF ELECTRICAL HEATING TO STEAM HEATING

There is a good scope to conserve energy if steam heating is used instead of electrical heating. A suitable design and scheme are suggested during a detailed energy audit study.

1.14.6 Energy Conservation in Steam Generation and Distribution

In a fertilizer plant, the steam generation and distribution network must work together. Many points must be followed for steam generation as well as for distribution by the operation personnel.

In the evaluation of boiler efficiency, the boiler efficiency is initially unknown. Testing per the standard has to be done in a detailed audit. Only then can energy conservation measures be suggested.

1.14.7 Steam Distribution System

There may be wide fluctuation in steam demand or no pressure control in the steam system.

1.14.8 Energy Conservation by Meticulous Water Resources Management

Raw water availability is another prime factor. Water can be conserved by:

1. Increasing cycles of concentration in cooling tower.
2. Recycling the water going into the drain.
3. Minimizing boiler blowdowns (too much blowdown leads to energy loss).
4. Recovering boiler blowdown water as a part of the makeup to the cooling tower.
5. Recovering final rinse water form cation and anion exchangers and feeding to cooling towers as a part of makeup water.

Perhaps no treatment is given to cooling water. Water balance of the complete system and the chemicals required for dosing are included in the detailed audit.

1.14.9 Energy Conservation in Ball Mill of Rock Phosphate

There is a good scope for conserving energy in the milling system, and the details are worked out in the final audit.

1.14.10 Energy Conservation in Furnace

By improvement of the present insulation

The furnace must be properly insulated. If insulation is poor, 30–40% of the heat is radiated through the walls to the atmosphere.

During the audit period, we measure the surface temperature to determine whether it is high (average of 83°C) and perhaps to suggest replacement of the present insulation material with better-quality material. The saving associated with and the investment required for proper insulation are included in the detailed audit.

By proper operation of the furnace

The design output of the furnace given by the vendor is based upon 100% utilization of the furnace. This assumes that the furnace is operated at full power without any idle period.

The operation of the furnace is a vital area, and a responsible supervisor has to monitor the operation so that wastage of energy is reduced.

A correct operating procedure is evolved in the final audit after watching some ten sessions of operations.

1.14.11 Energy Conservation by Using Additive in HSD and Furnace Oil

There is good scope to reduce the consumption of HS diesel and furnace oil by adding additive. The details of the additives to be used are given in the final audit.

1.14.12 Conservation on Capacity Utilization on Stream Factor

Capacity utilization has a strong impact on energy consumption in a DAP (diammonium phosphate) plant. The specific energy consumption is very high at 40% load, and as the plant load increases, there is a reduction of energy.

Another factor to be considered is the percentage of on-stream efficiency and the on-stream factor. Percentage on-stream efficiency is defined as the ratio of the percentage capacity utilization to on-stream factor. The higher the percentages are of on-stream efficiency, the lower the energy consumption will be.

If there is power tripping of, say, 80 hours in the past 10 months, this could be a reason for higher specific energy consumption of the plant. The remedial methodology is worked out in the detailed audit.

1.14.13 Energy Conservation by Continuous Feeding of Sulfur

If sulfur is dumped into a hopper by manual loading, by having a conveyor system to feed sulfur at steady rate, steam consumption can be reduced.

1.14.14 Energy Conservation in Agitators

There is good scope for energy conservation in agitators. The details are laid out in the detailed audit after carrying out measurements and tests.

1.14.15 Energy Conservation in Lighting System

The lighting of the plant is given in an annex. For example, the total lighting load of the plant 56 kW. The plant does not have a separate transformer or energy saver for the lighting system. There is no timer or photocell for switching off the lighting. All are manually controlled.

There is a very good scope in lighting. A detailed study has to be made on the present system of lighting, the illuminating level required in different sections, the replacement of unnecessary light, modifying the switching arrangement, the use of natural lighting wherever possible, etc. The illumination level is checked with a LUX meter and included in the detailed audit.

1.14.16 ENERGY OPTIMIZATION OF MOTORS

The load on each motor is measured during the preliminary survey and given in an annex. For example, an induction motor might operate at its maximum efficiency and develop a better power factor when loaded to above 70% of the rated capacity. Some motors might be underloaded and others overloaded.

A detailed study of the loading pattern of underloaded motors has to be done, taking into account the power factor at the load end. This can be done by connecting a demand analyzer at the end, which gives the actual kW, KVA, power factor, and current. This is all done in a detailed energy audit.

If there are no energy efficient motors in the plant, suitable suggestion are given in detailed energy audit.

1.14.17 ENERGY CONSERVATION BY INSTALLING SOFT STARTER

There is good potential to conserve energy by installing soft starter on some equipment. The actual savings can be confirmed only after carrying out measurements of the cyclic load of the motors.

1.14.18 ENERGY CONSERVATION BY REDUCING CABLE LOSSES

It might be found that there is wide variation in the voltages at the load end, which necessitates the installation of capacitors banks at the load end. The details are given during the final audit after carrying out feeder-wise power flow measurements. The low power factor increases transmission losses in the cables. The cable losses can be found by knowing their thicknesses, the distance of equipment from the main supply point, and the power flow measurements in the feeder.

1.14.19 ENERGY CONSERVATION BY LOAD MANAGEMENT

The economics of installing a DG (distributed generation) set are worked out in the detailed audit, as well as the balance of power between imported power and in-house generation. The cost of the imported power is escalating day by day. Hence every effort has to be made to reduce demand by increasing the load factor. Suitable suggestions are given in the detailed audit.

1.14.20 ENERGY CONSERVATION IN LUBE OIL ADDITIVES

We suggest using additive for the lubricating oil used in DG sets. This reduces maintenance problems and the consumption of the lubricating oil. The selection of the right additive and dosage is given in detailed audit.

1.14.21 ENERGY CONSERVATION IN PUMPS

An individual detailed study has to be done for each and every pump to decide the right conservation measure.

1.14.22 ENERGY CONSERVATION IN GEAR REDUCER

There is good scope for introducing helical gear reducers. The details are given in the detailed audit.

1.14.23 UTILIZATION OF SOLAR HEATING

The technical feasibility, investment, and payback for the installation of solar heaters are covered in the detailed audit.

1.14.24 ENERGY CONSERVATION IN WELDING SETS

Perhaps the total load of welding sets is 48 KVA, and out of 7 welding sets, 4 are working 8–12 hr/day. There is a good scope for optimizing energy consumption in welding sets. The detailed will be given in the detailed audit.

1.14.25 ENERGY CONSERVATION NORMS/MONITORING AND TARGETING

If the specific energy consumption of the plant based on the data furnished varies from, say, 20 to 45 units per tonne of production, then energy norms, both electrical and thermal, have to be worked out for each product. Energy norms enable XYZ pvt. Ltd to monitor and manage energy just like any other resources. It would also help to the organization reduce its present energy consumption.

Engineers can be trained in the monitoring and targeting of energy, which is very important to achieve success in energy conservation. Graphical techniques like the CUSUM technique using energy vs. production and SEC vs. production can be taught to the supervisory staff.

1.15 ENERGY CONSERVATION AWARENESS

The habits of energy conservation should be cultivated in the minds of all employees from the workers to the top management level. For this we suggest that higher officials should set an example by switching off the lights, fans, and air conditioners while leaving the office. These practices are followed by the officials in many industries. Posters and stickers for energy conservation may be pasted all around the plant.

1.15.1 ENERGY CONSERVATION CELL

We strongly recommend the formation of an energy conservation cell headed by a general manager (works) who will devote the time needed to energy conservation measures. The Auditor can also assist firms in the formation of the cell and in establishing other systems required for the initiation and implementation of the energy conservation program.

1.16 CONCLUSION

This chapter presented an outline of the energy auditing and accounting exercise of an electrical energy system from generation to distribution, the various losses involved in these energy transformation or/and transmission stages, and the details of reduction methods adaptable to India. To validate the viability of energy auditing, energy savings and related financial implications have been discussed. In addition, modern, scientific technologies being used and undergoing design and development in developed countries such as the United States, Canada, Switzerland, China, and elsewhere were also studied.

2 Energy Conservation Opportunities in Thermal Systems

L. Ashok Kumar and Gokul Ganesan

CONTENTS

DOI: 10.1201/9781003203810-2

LEARNING OUTCOMES

At the end of this chapter, the reader will be able to understand:

- Energy conservation opportunities in thermal systems.
- Technocommercial saving calculations for all energy saving proposals.
- Energy efficiency in boilers and steam distribution systems.
- Energy conservation opportunities in furnaces, waste heat recovery, cogeneration and cooling towers.

2.1 INTRODUCTION

Energy can neither be created nor be destroyed. It may be transformed from one form to another.

As stated in the law of conservation of energy, every industry consumes different types of energy as per their process requirement. Electricity is one of the major forms of energy consumed by all industries either directly for the process or indirectly for the utilities. Electricity is generated from different sources of energy irrespective of requirements. Other than electrical energy, other forms of energy, such as thermal energy and chemical energy, are also used in many industries. Thermal energy can be generated using fuels such as coal, diesel, petrol, biogas, etc.

Thermal systems in industries vary based on the process of a particular industry. A few common types of thermal energy equipment are:

- Boilers.
- Stream distribution systems.
- Heat exchangers.
- Furnaces
- Waste heat recovery systems.
- HVAC systems.
- Cooling towers.
- Insulation of thermal systems.

In this chapter, energy conservation opportunities in thermal systems in different industries are explained, along with the related technocommercial calculations.

2.2 ENERGY SAVING PROPOSALS (ESP) IN BOILERS

GENERAL RULES OF THUMB

1. An efficiency loss of 1% occurs with every 22°C increase in steam temperature.
2. Just 3 mm of soot can cause an increase in fuel consumption of 2.5%.
3. For every 20°C rise in combustion air temperatures, efficiency improves by 1%.
4. For every 15°C rise in feedwater temperature, efficiency improves by 3%.
5. For every 1% reduction in excess air, there is approximately 0.6% rise in efficiency.
6. For every 6°C rise in feedwater temperature, there is approximately 1% of fuel saving in the boiler.

$$\text{Boiler efficiency} = \frac{\text{Heat in steam output}\left(\text{Kcal}\right)}{\text{Heat in fuel input}\left(\text{Kcal}\right)} \times 100$$

$$\text{Evaporation ratio} = \frac{\text{Quantity of steam generation}}{\text{Quantity of fuel generation}}$$

$$\% \text{ of excess air} = \frac{O_{2\%}}{21 - O_{2\%}} \times 100$$

2.2.1 INCREASE STEAM OUTPUT PER KG OF FUEL BURNED THROUGH IMPROVED EFFICIENCY

Present Status

The efficiency of boiler varies from 65 to 75% depending on the calorific value of fuel. This can be increased to 82%.

Modification Proposed

A thorough study and design check of boilers and connected equipment indicate that there is scope for increasing the efficiency from existing 65–75% to 82%.

Energy Saving Potential and Benefits

The annual savings by increasing the efficiency of boiler by 10% is Rs. 52 lakhs.

Investment and Payback

Investment is Rs. 5 lakhs for design check and study and Rs. 30 to Rs. 40 lakhs for modification. The payback period is 1–1.5 years.

2.2.2 INSTALL ECONOMIZER IN BOILER FOR PREHEATING FEEDWATER

Present Status

There is no economizer for preheating the boiler feedwater in the boilers. The flue gas temperature is very high i.e., 300°C in boiler. A lot of heat is wasted due to high flue gas temperature. The present feedwater temperature is in the range of 60–70°C.

Modification Proposed

It is suggested to install economizer for boiler only to preheat the boiler feedwater to 120°C.

Energy Saving Potential and Benefits

There will be a 3% saving in fuel. The annual savings in fuel is Rs. 3.25 lakhs.

Investment and Payback Period

Investment for the economizer is Rs. 4 lakhs, and the payback period is 15 months.

2.2.3 IMPROVE THE EFFICIENCY OF THE BOILER BY SOOT BLOWING AND CONTROLLING EXCESS AIR

Present Status

The boiler efficiency was found to be 84% and can be increased to 86%.

The feedwater temperature is 110°C to the economizer, and the fuel is used to raise steam from this initial condition to 410°C superheated steam at 42 Kg/cm2 pressures. Two parameters that affect the efficiency of the boiler are excess air and high fuel gas temperature. The CO2 value of 14% and exit flue gas temperature of 186°C indicate that the reading is wrong.

Modification Proposed
The heat absorbing surfaces have to be cleaned frequently by means of soot blowing. The excess air level has to be monitored and controlled.

Energy Conservation Potential and Benefits
The annual savings by improving the efficiency of boiler is Rs. 16 lakhs.

Investment and Payback
The investment for repairing the soot blower and combustion controller, etc. is Rs. 12 lakhs. The payback period is 9 months.

2.2.4 INSTALLATION OF AIR PREHEATER TO BOILER

Present Status
At present there is no air preheater in the boiler. The combustion air is being delivered at ambient temperature to the burners.

Modification Proposed
It is suggested to install an air preheater to the boiler. The preheated air will be advantageous since it enhances the effectiveness of the combustion of oil by aiding atomization. Also, sensible heat to the air is saved to that extent, and the fuel consumption will be reduced.

2.2.5 CONVERT THE EXISTING GRATE-FIRED BOILER TO FLUIDIZED BED BOILER FOR IMPROVING EFFICIENCY OF BOILER

Present Status
At present the coal is burned on a travelling grate, and it is distributed on the grate by means of a mechanical spreader. There is a lot of unburned carbon in the grate. The boiler efficiency is 70–75% (max). There is a lot of excess air even though the CO_2 reading indicates an excess air ratio of 1.7 (max). The calorific value of coal also varies very much. The ash content is high around 40%.

Modification Proposed
The existing pressure parts arrangement should be maintained with the further addition of bed coils for providing a heat transfer area for raising steam, as well as for superheating and at the same time maintaining the bed temperature within 850°C. The existing travelling grate, spreader, and bottom ash discharge hopper, etc. should be removed, and the arrangement for the fluidized bed boiler should be done.

Energy Conservation Potential and Benefits

1. The combustion efficiency is increased to 95–98% by providing recirculation of the possible carryover of coal particles with fly ash. The bed coils keep the temperature of the bed always between 850 and 950°C to avoid fusion of ash and hence collapse of the bed. By converting the existing boiler into a fluidized bed boiler, the fly ash carryover is also reduced. As the low velocity of fuel gas is maintained in the portion above the grate, the ash carryover can be restricted.
2. There is also a possibility of increasing the boiler output by 5–10% depending on the present heat transfer area.
3. The annual savings obtained by the proposed modification works out to Rs. 37.92 lakhs.

Investment and Payback

Total investment for the proposed modification will be Rs. 45 lakhs, and it will be paid back in 15 months.

2.2.6 Usage of Fuel Oil Additive for LSHS

Present Status

At present, LSHS is used as fuel in the boiler and in the thermic fluid heater without any additive. The annual fuel consumption is 11,12,164 kg.

Modification Proposed

It is suggested to use fuel additive in the LSHS. There will be 1% of saving in fuel by using the fuel additive.

Advantages

1. Use of fuel oil additive will result in improved atomization, leading to better combustion.
2. The inhibitor in the fuel additive prevents corrosion.
3. Improved combustion leads to reduced fouling and soot formation.

Annual savings by adding fuel additive in the LSHS fuel oil is Rs. 84,524.

Investment and Payback Period

The investment for fuel additive is Rs. 47,143, and the payback period is 7 months.

2.2.7 Modify Burner in Boiler to Improve Combustion Efficiency

Present System

There is excess air in the flue gas as the burners have become defective.

Proposed System

It is suggested to replace the burner system with an energy efficient burning system in three boilers. The excess air fed to IAEC, W. W.NO. 1,2,3 boilers is 30–60%. This means that heat is carried away by the excess air fed to the boiler.

The efficiency of the boiler cannot be calculated by the direct method, since the steam flow reading appears to be incorrect. Hence the efficiency was calculated with the indirect method.

2.2.8 CONVERSION OF COAL-FIRED THERMIC FLUID HEATERS TO FLUIDIZED BED COMBUSTION HEATERS USING BAGASSE

Present Status

The thermic fluid heater available in the plant has a capacity of 4,00,000 kcal/hr coal is used as fuel. The thermic fluid flow rate is 26 m³/hr.

Modification Proposed

Convert the existing coal-fired thermic fluid heater to fluidized bed combustion using bed combustion of bagasse available from the nearby sugar factory.

Energy Conservation Potential and Benefits

1. The fluidized bed combustion system has the ability to burn bagasse and bagasse mixed husk (50%: 50%).
2. Fluidized beds permit variations in the moisture content of the fuel without impairing combustion.
3. The fluidized bed combustor is very compact in size and can be fitted in the available area.
4. The benefits include high efficiency, quick response to change in the load, no clinker formation, and relatively maintenance-free operation.
5. By adopting this modification, there will be a saving of Rs. 2.5 lakhs per year considering 50% capacity utilization.

Investment and Payback Period

The cost of investment for retrofitting for fluidized bed combustion is around Rs. 5 lakhs, which will be paid pack in just 24 months. This requires the preparation of bagasse suitable for operation. Hence a shredding machine is required in this case.

Alternately, we can offer a dump grate furnace for burning 100% bagasse or combustion of 50% bagasse and 50% husk at a cost of Rs. 3,75,000 with a payback period of 18 months. Compared to the fluidized bed furnace, this can be operated by even semi-skilled workers. This does not require shredding machines. The bagasse as received can be fed to the furnace. In both these cases, the bagasse handling is manual only.

2.3 ENERGY SAVING PROPOSALS IN STEAM SYSTEMS

2.3.1 PROVIDE SEPARATE LINE FOR CONDENSATE

Present Status

At present there is no condensate recovery. The steam distribution system was found disconnected. The amount of fuel used for steam generation can be reduced 10–30% by returning steam condensate.

Modification Proposed

We suggested using the condensate return to the boiler for use as feedwater. By donating this modification, the following can be achieved:

1. Saving of heat energy
2. Saving of makeup boiler feedwater
3. Saving of energy and chemical used in the water treatments operation
4. Reduction in water pollution

We also suggest insulating the condensate recovery steam. The saving obtained by providing a separate line for condensate recovery is Rs. 2,88,360.

2.3.2 Replace Existing Electrical Heating System with Steam Heating System for (1) LSHS Heating and (2) Air Heating in Process Section

Present Status

1. **LSHS heating:** In DG sets, LSHS is used as fuel. Two 24 kW electrical heaters are used for preheating the LSHS.
2. **Air heating in process section:** Presently a 72 kW electric heater is used in the process section to heat air to 180 °C. The cost of electrical energy is quite high. The average power cost, at the present pattern of in-house generation and imported power works out to Rs. 2,107/kWh.

The steam cost is almost nil since the steam is generated from waste heat boiler only.

Modification Proposed

Replace the existing electrical heating system by a steam heating system. Keep the present system as such for startup, emergencies, and fine-tuning.

Energy conservation potential and benefits: The annual savings in energy works out to 9,50,400 kWh costing Rs. 20 lakhs.

Investment and Payback

The estimated investment comes to Rs. 3 lakhs, and it will be paid back in just 2 months.

2.3.3 Improve the Efficiency of Steam Generation and Distribution System

Present Status

1. The efficiency of the boiler is less than 83%. There is good scope to improve that by 5%.
2. The insulation of the steam piping in the boiler surface is not proper.
3. Steam leaks are observed in many places. It is suggested to repair steam traps and arrest the steam leakage.

Modification Proposed

It is suggested to first carry out the insulation for 100 m2 wherever the temperature is 15°C above the ambient temperature. Subsequently the remaining surface of the boiler can also be reinsulated.

Energy Conservation Potential and Benefits

The annual saving by changing the insulation of boiler surface is Rs. 58,200.

Investment and Payback

The investment for changing the insulation is Rs. 1,00,000, and the payback period is 21 months.

2.3.4 IMPROVEMENT OF EFFICIENCY OF THE TURBINE

Present Status

1. The efficiency of the turbogenerator 2500 kW.
2. The specific steam consumption for the turbine is determined.
3. The steam consumed per kilowatt-hour of power generated by the turbine is 19.77 kg/kWh.
4. The specific steam consumption of the mill turbine is 3.96 tons/kWh.

Modification Proposed

The specific steam consumption of the turbines seems to be on the high side. The possibility of improvement in the efficiency of the turbines can only be ascertained after detailed study of the:

1. Turbine section.
2. Turbogenerator section.

2.3.5 STOP STEAM CONDENSATE PUMP

Present Status

The steam condensate at the steam trap outlet of the steam system is collected in a tank and pumped to the deaerator.

Modification Proposed

It is suggested that the steam condensate may be directly routed to the deaerator by utilizing the outlet pressure of the condensate at 11 kg/cm2, and the pump can be stopped.

Investment and Payback

Investment value is Rs. 1 lakh, and the payback period is 10 months.

2.3.6 ATTEND TO STEAM LEAKS IN STEAM LINES AND VALVES

Existing Setup

There are steam leaks in the steam lines, and values (for both high-pressure and low-pressure steam) in many places in the plant are as indicated, along with the estimated annual heat loss.

Modification Proposed

1. Attend to steam leaks in the identified areas aggressively by putting up a special task force.
2. Periodical inspection of steam lines and valves for steam leaks should be done, and defects should be rectified then and there.

Energy Savings and Benefits

The annual saving by plugging in the steam leaks in pipelines and values is Rs. 1,72,786 per year.

2.3.7 OPTIMIZATION OF STEAM UTILIZATION

Present Status

The steam produced in the boiler is being used in the boiler house (i.e., 20%), the rayon plant for drying, the VSF plant for drying, and the CS2 plant.

The total quantity of air to be preheated per hour works out to 3070*20/61,400 kg for 20–25% excess air, and the temperature rise is about 70–80°C. Further, in all the plants the steam used for drying is very high.

Modification Proposed

Modify the atomization system, air preheating system, and drier. A separate engineering study is required.

Energy Conservation Potential and Benefits

1. It is found by calculation that the heat required for drying the material in the rayon plant is just 1% of the total heat supplied by the steam. Hence there is lot of scope to improve the performance of the drier. Further it is found that 15 T (tonnes) of steam are supplied to the rayon plant. Out of this only 4*2.8 = 11.2 T of steam are consumed in the drier. The balance quantity of steam, i.e., 3.8 T/hr, remains unaccounted for. The insulation of the driver also can be improved by increasing the thickness of insulation so as to minimize energy loss through the walls.
2. By reducing the steam consumption by 10%, the quantity of steam consumed in the various auxiliaries can be reduced by tonnes, which can be used for running a VAM or meeting some other requirement.
3. The annual energy savings is Rs. 128 lakhs as given next.

Investment and Payback Period

The investment for the modification is Rs. 100 lakhs, and the repayment period is 10 months.

2.3.8 INSTALL THE RIGHT STEAM TRAPS

Present Status

20 thermodynamic steam traps are used, and they fail six months after installation.

Modification Proposed

It is suggested to replace defective thermodynamic traps with inverted bucket traps. Also purchase an ultrasonic leak detector for checking steam traps periodically.

Energy Conservation and Benefits

The annual savings is Rs. 1 lakhs, and the investment required is Rs. 1.6 lakhs, which will be paid back in 19 months.

2.4 ENERGY SAVING PROPOSALS IN FURNACES

2.4.1 FUEL EFFICIENCY IMPROVEMENT IN A GLASS FURNACE

Present Status

The present efficiency of the furnace is low.

Modification Proposed

1. It is suggested to work out the heat balance for the furnace.
2. Improvement in furnace efficiency can be achieved by using optimum excess air and reducing radiation loss by the application of insulation bricks or increasing the wall thickness.

2.4.2 MINIMIZE THE HEAT LOSS THROUGH THE EXHAUST FANS AND OPENINGS AT THE ENTRANCE AND EXIT

Present Status

There is no closure for the hot chambers in cleaning sections 1 and 2.

Modification Proposed

Provide doors in such a way that they will provide good sealing, preventing any air-flow into or from the ovens; spray cleaning chambers.

In the case of section 1, a sliding door operated by means of photoelectric cells may be provided. A double door system will provide minimum leakage.

If sliding doors are provided in section 2, they will open by sensing the component by means of a photocell. The doors' opening will be for entry of the chassis whose movement at that time can be made faster. Then there will be considerable savings. The exhauster and ventilators must be periodically opened just for 10 minutes in one hour's time and closed perfectly. Through this arrangement, we can avoid all possible heat losses through exhauster and ventilators and also the losses through openings. Air seals need not be provided. In such cases, the total heat loss during operation from all sections can be reduced by a minimum of 60%.

Energy Conservation Potential and Benefits

The estimated savings by minimizing the heat loss through the exhaust fans and openings at the entrance and exit are:

1. By providing manually operated airtight doors = Rs. 1,28,100.
2. By providing sensor-controlled airtight doors = Rs. 6,25,650.

Investment and Savings

The estimate for providing manually sealed doors and frame is Rs. 1,60,000, and for sensor-controlled doors, it is Rs. 6,00,000, which will be paid back in 15 and 12 months, respectively.

2.4.3 REPLACE PRESENT CONTROL PANELS OF RESISTANCE HEATING OVENS AND FURNACES BY THYRISTOR CONTROL PANEL

Present Status

The plant has four ovens.

The power to these furnaces is by a thyristor control system with suitable temperature sensors.

Energy Conservation Potential and Benefits

The thyristors are triggered on at different wave points by thermocouple sensors and temperature controllers. The power is controlled smoothly and precisely depending on the furnace temperature and heat requirement, eliminating temperature overshoot and undershoot and maintaining a stable process temperature.

1. The power to the heater is controlled precisely to have an optimum consumption, resulting in savings of about 5–10% of energy. The annual energy savings that can be realized works out to 40,536 kWh costing Rs. 2,00,248.
2. The useful life of the heating element is extended.
3. Reduced breakdown, reduced maintenance, and reduced element replacement mean consequently reduced cost.
4. Close tolerance of temperature, almost eliminating overshoots and undershoots, means that the quality of the product is improved.

Investment and Payback Period

The estimated total investment works out to Rs. 3.7 lakhs, which will be paid back in 22 months.

2.4.4 INTERLOCK OPERATION OF EXHAUST BLOWER WITH DOOR OPENINGS IN WDO

Present Status

During the detailed energy audit, operation of the fans in the ovens was studied in detail for possible energy savings.

The material is passed through the water dry-off oven, in order to remove its surface moisture.

Two exhaust blowers are present, respectively, at the inlet and exit of the WDO oven in order to avoid hot gas mixing with the ambient.

The total power consumption by the blower is 15.0 kW.

The water drier of the oven in the plant is not provided with a door. This allows the continuous escape of hot gas from the oven, resulting in energy loss.

There is good potential to reduce the power consumption of the existing blowers by providing pneumatic doors for the ovens and interlocking them with the blowers.

Modification Suggested

We recommend the following:

1. Install pneumatically operated doors for water dry-off oven.
2. Interlock the operation of the WDO blower with the door opening.

On a conservative basis, at least 60% of the power consumed by the blower can be reduced.

Investment and Payback Period

The annual energy saving potential is Rs. 1.1 lakhs. The investment required for doors in the WDO and interlocks would be Rs. 1.0 lakhs. The simple payback is 11 months.

2.4.5 Installation of Flux–Maxiox in Furnace Oil Supply Line

Present Status

There is no flux–maxiox system in fuel's supply line.

Modification Proposed

The flux–maxiox increases fuel efficiency and reduces exhaust emissions. The unit generates a strong enough flux field to substantially change the hydrocarbon molecule from its para state to the higher energized ortho state. This results in higher energy output for the same quantum of fuel used. And can be installed on the fuel supply line feeding the fuel from the fuel tank to the burner on the fuel pump discharge.

2.4.6 Use LPG as Fuel instead of Town Gas

Present Status

There is an imported furnace for heat treatment of 400 kg/hr. of components using town gas.

Modification Proposed

Use LPG as the fuel

1. LPG, being a clean fuel, will not pollute the environment. Sufficient care is to be taken to avoid any gas leakage through joints.
2. A temperature controller is also easy and almost acts as an electrical temperature controller.
3. The saving in energy cost is very high.

Energy Conservation Potential

Specific gross heating is taken as 750 kg/hr, and specific heat consumption is 0.442 kWh/kg for heating steel to a maximum temperature of 900°C. The heat required is 0.166 kWh/kg.

The actual specific heat consumption 10.442/0.166, which is 2.663 times higher than the theoretically required heat consumption. Hence there is scope to reduce this also.

The heat requirement is met by town gas whose calorific value is 3442–3713 kcal/Nm^3. The specific gravity of gas is 0.553 kg/ Nm^3 as an average.

Savings

If the same heat is to be supplied by electrical means, the total power equipment is 166.4 kW. This power may have to be higher for providing temperature control facilities and to take care of the thermal efficiency factor.

2.4.7 IMPROVE THE EFFICIENCY OF DRIER

Present Status

The company has one drier of capacity 500 kg/hr. HSD is used as fuel in the drier. The drying time is 2 hours, and the temperature maintained is 104–108°C. The average fuel consumption in the drier is 47.5 L/T.

DURING OUR AUDIT, WE OBSERVED THE FOLLOWING:

1. There is a crack in the front end of the furnace through which hot gas is escaping.
2. The surface temperature measured was high.
3. There is leakage through the joints in the drier.

Modification Proposed

1. It is suggested to arrest the leakage through the crack and joints.
2. Change the existing insulation of the drier, which should be only 15°C above the ambient temperature.
3. Order six more trolleys to increase the ripping time so that rubber entering the drier is in a dry state.
4. Use fuel oil additive Micromx 1000 D to improve combustion, fouling, and soot formation.

Energy Conservation Potential and Benefit

These measures will reduce fuel consumption by 8 L/T.

The annual energy savings by proper insulation and arresting the leaks is Rs. 1,86,559.

Investment and Payback

The investment for improving insulation and arresting leaks is Rs. 1,00,000, and the payback period is 6 months.

2.4.8 Keep Usage of Holding Furnace to the Minimum Possible to Avoid Energy Loss

Present Status

It is always better to tap the metal from the induction melting furnace and pour it into molds. In this case, the holding capacity of each furnace is 10 T, and the melting capacities of the furnaces are of 2 T and 5 T, respectively. The holding furnace will be unnecessarily holding molten metal, and the heat loss through the cooling water and the top surface to the air will be high. Further if we are to hold the melt in this furnace to this full capacity, we have to wait for at least 2 melts to fill this furnace to this full capacity, and so we have to wait for at least 2 melts at any one of the furnaces, which takes around 1.5 hours for melting.

Modification Proposed

It is suggested to use the holding furnace to the minimum possible to avoid energy loss. By using the holding furnace, we will be wasting energy in addition to increasing the handling of molten metal. It will be worthwhile if the melting rate of furnace is higher and the pouring rate of the casting is lower. The surplus material may be stored in the holding furnace to avoid wastage of energy in melting.

Energy Conservation Potential and Benefits

The annual savings obtained by the usage of the holding furnace is Rs. 13,40,510.

2.4.9 Shot-Blast the Returns to Remove the Green Sand

Present Status

The runners and risers from the molding are covered with green sand and are remelted, resulting in slag formation. The metal loss is up to 8%.

Modification Proposed

It is suggested to shot-blast the return before feeding into the furnace. A minimum 50% slag formation can be avoided.

Investment and Payback Period

At least 30% reduction in power consumption; i.e., 6 units can be achieved.

2.4.10 Install Ladle Preheating for Furnace

Present Status

The furnace molten metal tapping temperature = 1550°C.
There is a temperature drop of 90°C from the tapping point to the first pour.

Modification Proposed

It is suggested to install an oil-fired furnace for preheating the ladle and to practice ladle preheating before tapping the molten metal.

Investment and Payback Period

A minimum of 1% savings can be achieved by avoiding the drop in temperature of 90°C, and the payback period is 7 months.

2.4.11 LOAD THE FURNACE TO THE FULL CAPACITY

Present Status

The melting furnace is loaded only to 65%. The reason is that different types of components are used for melting.

Modification Proposed

It is suggested to increase the loading of furnaces by providing additional tiers. The fabrication of new fixtures are required for additional tiers.

Investment and Payment Period

Investment cost is Rs. 25,000, and the simple payback period 5 months.

2.4.12 RESCHEDULE THE BATCH OPERATIONS OF FURNACE

Present Status

Each furnace batch takes 3 hours from cold start to the batch.

Modification Proposed

It is suggested to do three batches one after other in the same shift itself. By this, 20 minutes of heating are saved.

2.4.13 REDUCE THE TAPPING TEMPERATURE

Present Status

The pouring temperature of metal ranges from 1360 to 1425°C. The metal is melted at a higher temperature to take care of deficiencies in ladle heating, temperature drop due to delay in pouring, logistics, etc.

Modification Proposed

It is suggested to maintain the pouring temperature at 1400°C.

Energy Conservation Potential and Saving

Every 10°C rise consumes 10 units/heat.

Investment and Payback Period

Investment is nil.

2.4.14 MINIMIZE POURING TIME

Present Status

In the molding machine, the pouring of metal is done by hoist. The present pouring time is 40 minutes. Energy consumption during the pouring time is 40 kWh.

Modification Proposed

1. By properly planning the infrastructure like space and the availability of the hoist, the pouring time can be reduced to 30 minutes.
2. Fill up the ladle to full capacity.

Investment and Payback Period

Annual savings are Rs. 75900, along with an annual increase in production by 300 MT (metric tonnes).

2.5 ENERGY SAVING PROPOSALS IN INSULATION AND REFRACTORIES

2.5.1 INSULATE THE PIPELINE AND EQUIPMENT TO REDUCE HEAT LOSS

Present Status

In most outdoor piping and equipment, the insulation is damaged and in many places wet. This causes heightened heat loss, resulting in greater consumption of steam or temperature drop of the fluids. The replacement of insulation is to be done in such a manner that no water gets into the insulation.

Modification Proposed

It is suggested to completely redo the insulation using a better method for avoiding moisture ingress and sealing the ends to prevent entry of rainwater.

Energy Conservation Potential and Benefits

In the first phase, the area for replacement of insulation is taken as 500 m².

The heat loss through the bar surface that needs insulation, as well as through the poorly or noninsulated pipelines, equipment, and vessels, works out to 500–600 kcal. This can be brought down to 150–200 kcal/m²/hr. The annual savings by improving insulation is Rs. 6,57,280.

Investment and Payback Period

The investment for improving the insulation is Rs. 3,00,000, and the payback period is 6 months.

2.5.2 REPLACE THE EXISTING REFRACTORY AND INSULATION WITH CERAMIC WOOL BACKED BY LRB MATTRESS IN FURNACE

Present Status

At present the furnace is lined with refractory bricks. The furnace is heated up to 850°C, and there is heat loss through the refractory and insulation.

Modification Proposed

It is suggested to replace the existing refractory and insulation with ceramic wool backed by LRB mattress.

Energy Conservation Potential and Benefits

1. The ceramic fibers are alumino-silicate material in fiber form.
2. The advantage of ceramic fibers is low heat storage, lightweight construction, and resilience, as well as low thermal expansion, ease of handling, and repair and maximum operating temperature.
3. It can be used on any type of hot surface insulation in direct contact with aluminum, lead, zinc, copper, and alloys of these materials.
4. Due to low thermal conductivity and low heat storage, the furnace lined with ceramic fiber shows faster response and hence more accurate temperature control. This also results in a more uniform temperature distribution throughout the furnace.
5. The annual energy saved by providing ceramic wool backed by rock wool slab insulation is worth Rs. 3,32,790.

Investment and Payback

The estimated investment is Rs. 1,25,500, which will be paid back in 5 months.

2.5.3 Provide Ceramic Wool Lining Backed by Rock Wool Bonded Slab in Place of Refractory Bricks for Rotary Retort Furnace

Present Status

The retort is rotating inside a housing where the heater is located and the housing is lined. The temperature of the furnace is maintained at 850°C, and the surface temperature as measured is 116°C.

Modification Proposed

It is suggested to provide ceramic wool lining backed by rock wool bonded slab in the place of insulation bricks.

Energy Conservation Potential and Benefits

1. It can be used on any type of hot face insulation at the maximum temperature range of 1000–1600°C.
2. The annual energy saved by providing ceramic wool backed by rock wool slab insulation is worth Rs. 2,66,230.
3. It is not wetted by molten metals.

Investment and Payback

The estimated investment is Rs. 1,11,000, which will be paid back in 6 months.

2.5.4 Insulating the Uninsulated Pipelines and Changing the Damaged Insulation in Pipelines to Reduce the Heat Loss

Present Status

The insulation of steam piping and boilers is not properly done in some places, and insulation is missing, which results in higher surface temperature and heat loss.

Modification Proposed

The existing insulation of boilers is to be changed so that the surface temperature is only 15°C above the ambient temperature. At present the average surface temperature is 60°C when the ambient temperature is 25.4°C. Also insulate the uninsulated pipelines to bring down heat loss. The surface temperature of these pipes is found to be 100–130°C.

Energy Conservation Potential and Benefits

The annual energy savings by providing insulation materials are worth Rs. 1.34 lakhs.

2.5.5 Installation of Polypropylene Balls in the Electroplating Baths

Present Status

In the electroplating bath, the radiation losses are greater from the liquid bath since it is left open. The temperature in the bath varies from 50 to 80°C.

Modification Proposed

It is suggested to use polypropylene balls in the electroplating baths to avoid the radiation losses and also reduce losses from the liquid's surface.

The cost of polypropylene balls is Rs. 10,000, and payback period is 2 months.

2.5.6 Reduction of Heat Loss through Surface Insulation

Minimize heat loss through insulated surfaces and uninsulated valves, flanges, expansion tanks, and some pipes up to the surface of the ovens and their connected equipment.

Present Status

At present, valves and flanges are not insulated.

Modification Proposed

1. All the connected thermic fluid lines should be provided with additional 50 mm thick insulation.
2. The Thermo Pac surface should be insulated with 50 mm thick insulation.
3. The valves, flanges, and the expanded tanks provided in the thermic fluid heating systems should be insulated with 100 mm thick insulation.
4. The outer surface should be given a hard settings compound finish to minimize investment.

Energy Conservation Potential and Benefits

The estimates savings obtained by minimizing the heat loss through insulated and uninsulated surface works out to Rs. 2,15,370.

Investment and Payback Period

The estimated investment works out to RS. 322,000, which will be paid back in 18 months.

2.5.7 PROVIDING LIDS MODE OUT OF THERMOCOLE JUST ON THE SURFACE OF THE HOT LIQUID BATHS SO THAT HEAT LOSS THROUGH EVAPORATION IS AVOIDED AT THE END OF SHIFT

Present Status

There is no lid on the hot liquid baths, resulting in evaporation loss from all the heated baths of the treatment section.

Modification Proposed

A thermocouple (expanded polystyrene) may be made to a thickness of 75 mm to fit the inner size of the tank and be allowed to rest on the liquid surface, covering the free surface of the liquid.

Energy Conversation Potential and Benefits

The estimated savings obtained by providing lids made out of thermocouple works out to be Rs. 66.900.

Investment and Payback Period

The estimated investment works out to be Rs. 10,000, which will be paid back just 2 months.

2.5.8 INSULATE THE BARE STEAM LINES IN THE IDENTIFIED AREAS

Present Status

It was observed that some process equipment and steam lines are not insulated. There is heat loss through these uninsulated areas.

Bare steam pipeline, flanges, and hot process equipment give off heat to the atmosphere by radiation.

Modification Proposed

It is recommended to provide proper insulation.

Energy Conservation Potential and Benefits

1. Insulation reduces the heat loss due to radiation.
2. Safety of the workers is improved
3. The annual energy savings by insulation is Rs. 40,300.

Investment and Payback Period

The investment for insulation is Rs. 20,000 and payback period is 6 months.

2.5.9 INSTALL COVERS FOR TANKS WITH HEATERS

Present Status

Since there is no cover for the surface treatment tanks, heat is lost to the atmosphere.

Modification Proposed

Install insulated covers for all tanks with heaters.

Energy Conservation Potential and Benefits

The annual savings that can be achieved by implementing this proposal is Rs. 3.59 lakhs. The investment required is Rs. 1.30 lakhs, which is paid back in 5 months.

2.5.10 INSULATE FURNACE WALL

Present Status

The furnace walls are not insulated.

Modification Proposed

It is suggested to insulate the furnace wall to reduce the heat loss.

Energy Conservation Potential and Benefits

The heat loss is reduced. The estimated annual savings is Rs. 21,128 with an investment of Rs. 20,000, which will be paid back in 11 months.

2.5.11 PROVISION OF LID FOR 2 T FURNACE AND REPLACE EXISTING LID WITH CERAMIC WOOL IN 5 T FURNACE

Present Status

The 2 T furnace has no lid. So the heat loss is more through the opening of the furnace. The 5 T furnace lid is provided only with a castable lining, resulting in the heat loss through the lid.

Modification Proposed

It is suggested to provide ceramic-wool-lined lid for the 2 T induction furnace and also to replace the existing castable lining with a ceramic wool lining for the 5 T furnace.

Energy Conservation Potential and Benefits

1. The ceramic fibers are alumino-silicate material in fiber form.
2. The advantages of ceramic fiber are its low heat transmission, low heat storage, lightweight construction, resilience, low thermal expansion, ease of handling and repair, and maximum operating temperature.
3. It can be used on any type of hot face insulation in the maximum temperature range of 1000–1600°C.
4. It is wetted by molten metal and can be used in direct contact with aluminum, lead, zinc, copper, and alloys of these materials.
5. Due to low thermal conductivity and low heat storage, the furnace lined with ceramic fiber shows faster response and hence more accurate temperature control. This also result in a more uniform temperature distribution throughout the furnace.
6. The annual energy saved by providing ceramic wool insulation is Rs. 4,00,000.

Investment and Payback Period

The estimated investment is Rs. 64,800, which will be paid back in 2 months.

2.5.12 INSULATE THE BARRELS OF ALL THE ELECTRICAL HEATER IN LINES A, B, C

Present Status

At present the barrels of all the heaters are not effectively insulated. Surface temperatures as high as 296°C are recorded.

Modification Proposed

Arrest the radiation heat from the surface of the heaters of all the lines by providing split flange coupling with holes for thermocouples. The packing in the split flange will be ceramic fiber.

Investment and Payback Period

The estimated investment is Rs. 60,000, and the payback period is 2 months.

2.6 ENERGY SAVING PROPOSALS IN COGENERATION

2.6.1 REPLACE THE PRESENT CONDENSING TYPE OF TURBINE TO EXTRACTION TURBINE SO THAT THE PROCESS STEAM REQUIRED CAN BE TAPPED AT THE APPROPRIATE STAGE AS WELL AS BALANCE STEAM TO THE OPERATE LP TURBINE

Present Status

At present a portion of steam generated (18T/hr boiler) is being used to run the turbine generator, which consumes 13 T/hr of steam per hour for the generation of 2 MW. The balance steam is used in the process by reducing the pressure in the Pressure Reducing and Desuperheating Station (PRDS), and demineralized water is injected into steam to bring down the temperature for meeting the process requirement. A lot of heat energy is wasted in this arrangement.

Modification Proposed

The existing condensing turbine may be replaced by extraction using the total steam for the high-pressure turbine and tapping out steam at 12 kg/cm2 pressure, maintaining a flow rate of the 23 T/hr required for process. The balance 13 T/hr may be sent for power generation in the low-pressure turbine.

Energy Conservation Potential and Benefits

A preliminary calculation using an extraction type of turbine for power generation and for process requirement shows that there is a need to generate steam at 29 T/hr, which gives an output of 3.5 MW in the turbine (the existing output), and also the steam for process.

Hence by the proper selection of the turbine, an additional 1.5 MW of power can be generated by using the steam from available boilers.

The annual saving obtained by installing an extraction turbine works out to Rs. 4.32 crores.

Investment and Payback Period

Investment for the project will be Rs. 2.5 crores, which will be paid back in 7 months.

2.7 ENERGY SAVING PROPOSALS IN WASTE HEAT RECOVERY

2.7.1 UTILIZE THE HEAT GENERATED THROUGH INCINERATION OF THE WASTE AVAILABLE FOR RUNNING THE LOW-CAPACITY BOILER TO REDUCE OIL CONSUMPTION

Present Status

The waste generated in the factory is being disposed of, and there is no proper incinerator for destroying it.

Modification Proposed

The waste available and generated in the factory can be incinerated, and the hot gas can be utilized for running the low capacity turbine.

Energy Conservation Potential and Benefits

The annual savings by utilizing heat generated through incineration of the waste is Rs. 1,17,010.

Investment and Payback Period

The investment for installing the incinerator is Rs. 3,00,000, and the payback period is 31 months.

2.7.2 OPERATE THE EXISTING HEAT RECOVERY SYSTEM IN POT CONCENTRATOR FURNACES FOR RECOVERING THE WASTE HEAT

Present Status

The furnace is provided with an air heater, but it is not being operated. The oil consumption of 185 L/hr for increasing the concentration of sulfuric acid appears to be very high as the output of sulfuric acid is 25 T/day only. The existing gas temperature is above 250°C, and there is a scope for recovering the heat from this.

Modification Proposed

Repair the existing air heater for recovering the waste heat from flue gas and put it online.

Energy Conservation Potential and Benefits

The annual savings in oil by repairing the air heater and putting it online is Rs. 7,20,800 as given in an annex.

Investment and Payback Period

The investment for repairing the air heater is Rs. 1,00,000 for three furnaces, and the payback period is 2 months.

2.8 ENERGY SAVING PROPOSALS IN THERMIC FLUID HEATERS

2.8.1 IMPROVE COMBUSTION EFFICIENCY BY REDUCING THE EXCESS AIR IN THE EXIT GAS

To reduce excess air in the exit gas to improve combustion efficiency of both the thermic fluid heaters.

Present Status

At present, two numbers of thermic heaters are available in the plant. The capacity of the thermic fluid heaters is 10 lakhs kcal/hr for heater no. 1 and 6 lakhs kcal/hr for heater no. 2. The actual furnace's oil consumption for heater no. 1 is 110 L/hr, and for heater no. 2, it is 63.3 L/hr.

Modification Proposed

1. It is recommended to maintain the following parameters for both thermic fluid heaters.
2. Use a portable flue gas meter capable of measuring O_2 and CO_2 combustibles, etc. to analyze flue gas.
3. Operate the thermic fluid heater at 2–3% O_2 without combustibles in the exhaust gas, which is practically possible.
4. Conduct periodic refresher courses on fuel efficiency for the thermic fluid heater operator.

Energy Conservation Potential and Benefits

1. Combustion of a fuel is a chemical reaction. Combustion is the rapid combination of oxygen with fuel, resulting in the release of heat. The exact quantity of oxygen required to complete this chemical reaction of combustion is called the stoichiometric quantity of oxygen. The oxygen is usually derived from air. Combustion does not start in air or continue spontaneously below a certain temperature known as the ignition temperature of fuel with air.
2. It is essential to supply sufficient air to burn fuel completely. This sufficient quantity of air is always in excess of stoichiometric quantity. But too much excess air also takes a considerable quantity of heat released from the fuel since air contains 79% of inter nitrogen. The excess air supplied is determined by measuring the O_2 content or CO_2 content in fuel gas during our audit period.
3. The estimated savings obtained by reducing the excess air in the existing gas is around Rs. 2,19,700.

Investment and Payback Period

The investment for the fuel efficiency monitor is around Rs. 50,000, which will be paid back in just 3 months.

2.8.2 Conversion of Electrically Heated Water Baths in the Temperature Section to Thermic Fluid Heated Equipment

Present Status

The total capacity of the installed heaters in all the three electrically heated baths is 52.5 kW.

Modification Proposed

All three electrically heated baths shall be provided with thermic fluid heating coils with all suitable controls, and they will supply the required heat as they have surplus to meet this requirement.

Energy Conservation Potential and Benefits

1. The present electrical heater is consuming totally 52.5 kW in 1 hr, and its thermal equivalent is 45,150 kcal.
2. The estimated savings obtained by conversion of electrically heated water baths to thermic fluid heating works out to Rs. 5,58,240.

Investment and Payback Period

The estimated investment works out to Rs. 4,50,000, which will be paid back in 10 months.

2.8.3 Performance Improvement of Thermic Fluid Heater

Present Status

The efficiency of the thermic heater is 80.02%, whereas the design efficiency is 87%.

Modification Proposed

The efficiency of the heater can be improved by:

- Reducing the excess air used for combustion of fuel.
- Descaling the tubes during shutdown periods.

It is recommended to maintain the following parameters in the thermic fluid heater.

Energy Conservation Potential and Benefits

It is essential to supply sufficient air to burn the fuel completely. The quantity of air is always shown to be in excess of just sufficient. This takes a considerable quantity of heat released from the fuel since air contains some percentage of inherent nitrogen. The excess air supplied was determined by measuring O2 the content in fuel gas.

The estimated energy savings by reducing the excess air in the exits gas is around Rs. 61,060.

Investment and Payback

No investment is required.

2.8.4 Conversion of Electrical Heating to Thermic Fluid Heating

Present Status

At present, the plant has five 18 kW and two 36 kW heaters.

Modification Proposed

The heater in the plant may be modified for using thermic fluid heating. They shall be provided with temperature controllers. It is recommended to use a 2 lakh kcal thermic fluid heater.

The total (thermic fluid heater) full load for both plants is 1,06,210 kcal.

Energy Conservation Potential and Benefits

The estimated saving obtained by conversion of an electrical heaters to thermic fluid heater works out to Rs. 11.36 lakhs.

Investment and Payback

The estimated investment is Rs. 7 lakhs, and the payback period is 8 months.

2.8.5 Installing a Boiler Working on Thermic Fluid Heating for Quick Start-up in the Event of Stoppage of the Regular Boiler

Present Status

At present the standby boiler is always kept at the rated outlet pressure, and this is achieved by operating the burner as and when the pressure falls. This is consuming quite a huge quantity of oil. Approximately 14,080 kg of oil are consumed per month.

Modification Proposed

It is proposed to install a thermic fluid heated boiler that will always be kept under the rated pressure. The size of this boiler will be comparatively smaller. The thermic fluid heater will provide adequately by this boiler so that, in the event of failure of the regular running boiler, this boiler will automatically come online. Since the quantity of thermic fluid will be fully utilized from the entire thermic fluid heater and as the water quantity of this boiler is kept small, steam saving will be within 10 minutes from the start.

Energy Conservation Potential and Benefits

The estimated savings obtained by providing thermic fluid heating for the start-up boiler works out to Rs. 14,87,632.

Investment and Payback Period

The cost of the boiler works out to Rs. 20,00,000, which will be paid back in 16 months.

2.8.6 IMPROVE THE COMBUSTION EFFICIENCY OF AQUA THERM

Present Status

The capacity of the aqua therm is 6 lakhs kcal/hr. The total fuel consumption by the aqua therm is about 400 L/day.

The combustion analysis is 02–5.4% against the design 3%.

Modification Proposed

Monitor the O_2% in the fuel gas on a regular basis and to maintain an oxygen level of 3–4%.

Energy Conservation Potential and Benefits

There will be an approximate increase of 0.5% in the combustion efficiency if O_2 is monitored.

The total annual savings will be Rs. 9300 without any investment.

2.8.7 OPTIMIZATION OF THERMIC FLUID HEATERS

Present Status

At present two thermic fluid heaters are available in the plant, one for line II (Thermo Pac at a capacity of 6.3 lakhs kcal/hr) and one for line III (Sharptherm thermic fluid heater at a capacity of 15lakhs kcal/hr).

Modification Proposed

From the loading data, it is found that the Sharptherm thermic fluid heater is loaded to 60%, and the Thermo Pac thermic fluid heater is loaded to around 65%. Hence, it is suggested to run only the Sharptherm thermic fluid heater while both lines are running.

Energy Conservation Potential and Benefits

The savings obtained by the preceding modification is around Rs. 3,13,503.

Investment and Payback Period

The investment for this modification system is nil since the interconnections are already available and hence the payback is immediate.

2.9 ENERGY SAVING PROPOSALS IN COOLING TOWERS

2.9.1 REPLACE EXISTING METAL BLADES BY FRP BLADES FOR COOLING TOWER

Present Status

1. The location is the cooling plant and cooling tower.
2. The specification is 200 TR (tons of refrigeration) capacity

3. Fan motor rated power (kW) is 11.5.
4. Actual power kW is 5.93.

Modification Proposed

Replace the aluminum blades by new energy efficient FED blades. Using FRP blades means a minimum saving of 10% in energy.

Energy Conservation Potential and Benefits

The FRP blades are lighter than aluminum blades and do not allow material buildup. Hence the load on the fan is reduced, thus reducing power consumption. In addition:

1. These blades ensure easy maintenance and thus less downtime.
2. Due to the light weight of the blades, the size of the hub is reduced considerably, which provides a more effective air discharge area, resulting in more airflow with less pitch angle.
3. These blades also increase the life of a mechanical drive arrangement.
4. The $I_2 R$ loss on cables is reduced due to lower power consumption.
5. They are not subjected to corrosion or erosion, and fittings and material buildup is prevented; hence it maintains its aerofoil design. The estimated energy saved per annum is 5052 kWh, costing Rs. 17,834.

Investment and Payback Period

The investment for replacing metal blades by FRP blades is around Rs. 10,000, which is paid back just in 7 months.

2.9.2 Install Automatic Temperature Controller in the Cooling Tower System

Present System

The cooling tower fans are working continuously irrespectively of inlet water temperature.

The performance of cooling depends on ambient wet bulb temperature (which varies between day and night). Further, the cooling tower water temperature also depends on the cooling load.

The cooling tower fan capacity is designed for a maximum cooling load and the worst ambient conditions. Consequently, during the period of favorable ambient conditions (night time) and low cooling load, the cooling tower fan capacity is underutilized, resulting in wastage of energy.

The temperature is controlled manually.

Modification Proposed

Install automatic temperature controllers for cooling towers (28–30°C).

Energy Conservation Potential and Benefits

The controller switches off the fan when the cold well temperature goes to the set temperature and switches on when the temperature goes above the set temperature (28–30°C).

The savings obtained by installation of automatic temperature controller is around Rs. 1,69,294.

Investment and Payback

The estimated cost of installation of automatic temperature controller is Rs. 50,000, which will be paid just in 4 months.

2.9.3 CENTRALIZATION OF UTILIZE AND PROVIDING A COMMON CHILLED WATER LINE, CHILLED BRINE LINE, COOLING WATER LINE AND COMPRESSED AIR LINE

Present Status

At present, all the utilities are located at different places supplying cooling water, chilled water, and chilled brine solution for the various processes. There are also many small cooling towers with spare pumps. Locating all utilities at a central place for easy distribution and maintenance is a worthwhile proposition.

Modification Proposed

The cooling tower located at various place shall be clubbed, and there will be four cooling towers, each of 1000 TR capacity with two running at all times and two as standby. All the cooling towers have interconnections, and any cooling tower can be brought into the stream or taken out to suit convenience. All the piping will be common. Existing pipelines shall be made use of. All the chilled water plants and brine chilling plants shall be installed adjacent to the cooling tower, and there will be a central header from where the pipelines to various usage points shall be run. All the existing pipelines in the various buildings will be utilized and they will not be distributed. The air compressors also shall be brought under this roof so that there will be a centralized supply of compressed air.

Energy Conservation Potential and Benefits

All the utilities can be made to come online or be detached depending on requirements so that the running of the utilities shall be economical. Since there will be a central supply, the individual utilize shall be loaded to their maximum, and only when there is a further requirement will the other available unit be brought into the stream. The arrangement is aimed to achieve maximum utilization of all utilities in a very economical way. The loading analyzed indicated that 40% of the installed capacity is utilized. Hence the utilities can be run in such a way as to bring in savings.

The proposed arrangement is expected to bring in a saving of 1000 units in a day. Hence a saving of Rs. 1000 * 3.52 * 30 = Rs. 1,05,000 per month can be achieved. Yearly saving will be Rs. 12,67,200.

The modification will involve dismantling and shifting the equipment to a centralized place. The cooling towers may be disposed of, and the new unit proposed may be purchased and installed.

Investment and Payback Period

The total cost of modification will be around Rs. 20 lakhs, which can be paid back in 19 months.

2.9.4 INSTALL SIDE STEAM FILTER FOR COOLING TOWER OF POWER PLANT

Present Status

The water for the river is very dirty (total suspended solids [TSS] going 25 times more than normal), and it forms a coating on the condenser, resulting in drop in vacuum and heat transfer.

Modification Proposed

Install four side steam filters of 120 m3/hr each so that three filters take care of the fuel load when the filter is on backwash.

Energy Conservation Benefits

1. Long life of condenser lutes
2. Maintenance of the vacuum in condenser
3. Good heat transfer

Circulating Water

Management of the circulating water system is very important for achieving the desired vacuum, and a drop in vacuum causes efficiency loss of the turbine, resulting in a heavy loss of energy in the process of generating electricity.

Investment and Payback Period

The investment for four filters is 6 lakhs, which will be paid back in 12 months. The saving expected is 6 lakhs.

2.10 ENERGY SAVING PROPOSALS IN WATER TREATMENT

2.10.1 REVAMP THE EXISTING COOLING WATER TREATMENT

Present Status

Presently, zinc-based chemicals and chlorine are dosed in the cooling water system. The process water and cooling water parameters are given.

As seen from the algae formation and from the maintenance of the parameters, the cooling water program is not doing well. The present cycle of concentration is maintained between 1.5 and 2 in order to play it safe.

Modification Proposed

It is suggested to change the present system to zinc organophosphate-based treatment for scale and corrosion control with a biocontrol program for improving the efficiency of chlorine.

Energy Water Conservation Potential and Benefits

1. The cycle of concentration can be increased to 4 from 2. By increasing the cycle of concentration to 4, there is a possibility to save 50% of the present make-up water, i.e., 50% of 76 m³/hr of water = 38 m³/hr. There will be annual savings of Rs. 18 lakhs in the water bill by this.
2. Since water consumption in the plant comes down by 910 m³/day. The company can manage with 2790 m³/day itself, and it increases the storage capacity of the plant by one more day.
3. Heat transfer efficiency will improve.
4. There will be less fouling.

Investment and Payback

It is estimated that the proposed system will cost the same as the present system with better control. Hence the payback period is negligible.

3 Energy Efficiency in HVAC and Refrigeration Systems

Kishen Singh and Gokul Rajendran

CONTENTS

DOI: 10.1201/9781003203810-3

LEARNING OUTCOMES

At the end of this chapter, the reader will be able to understand:

- Fundamentals of HVAC and refrigeration systems.
- Various types of refrigeration systems used in industries.
- Energy conservation opportunities in HVAC and refrigeration.
- Real-time industrial case studies with technocommercial saving calculations.

3.1 INTRODUCTION

In the day-to-day work lifestyle, the atmospheric and ambient conditions greatly favor the effectiveness of the worker. In such a case, working conditions have to be considered with utmost care. It is where the HVAC (heating, ventilation, and air Conditioning) comes into picture. It will follow all the three thermodynamic laws to ensure maintaining favorable conditions. The process where the ambient condition is below 0°C, the temperature needs to be raised so that heating is the operation done through the heat pumps. Ventilation is the process of providing vents for the existing gases to the atmosphere and also inducing a fresh charge of air into the subjective space. Ventilation is provided naturally through density air movement (the thermosyphon effect) and is also forced airflow using exhaust fans and blowers. Air conditioning is the optimization of temperature below the ambient conditions to maintain human comfort.

3.1.1 REFRIGERATION AND AIR CONDITIONING

Refrigeration is the process of reducing temperature of the space or system to preserve the required conditions. It is used in the areas of food preservation and storage, production of ice, and pharmaceutical drugs. Air conditioning is the process of maintaining the temperature at 5–10°C below the ambient condition by altering the moisture level with respect to humidity to maintain the human comfort conditions. Generally, the human comfort condition is between 20 and 26°C.

3.2 CLASSIFICATION OF HVAC SYSTEMS

The different types of refrigeration systems are as follows:

BASED ON CYCLE OF OPERATION

- Vapor compression refrigeration system (VCRS)
- Vapor absorption refrigeration system (VARS)

BASED ON THE COOLING METHODS OF HEAT DISSIPATION

- Air-cooled system
- Water-cooled system

BASED ON THE COOLING MEDIUM

- DX (direct expansion) system
- Chilled water system

The different air conditioning systems are as follows:

BASED ON UNITARY OPERATION

- Room air conditioners
  - Window AC
  - High wall split AC
  - Cassette
  - Inverter AC
- Centralized air conditioners
  - Constant flow
    - Air cooled
      - Ducted split
      - Ceiling concealed ducts
      - Ceiling exposed ducts
      - Single packaged
      - Floor standing
    - Water cooled
      - Single packaged
      - Floor standing
  - VRF (variable refrigerant flow)
    - Air cooled
    - Water cooled
      - Ducted split
      - Concealed ducts
      - Enclosed ducts
      - Ceiling cassette
      - AHU (air handling unit)

BASED ON CHILLED WATER AND BRINE SYSTEMS

- Centralized air conditioners
  - Air cooled

- Water cooled
  - AHU (air handling unit)
  - FCU (fan coil unit)
  - Chilled beams
  - Underfloor heating

3.3 VAPOR COMPRESSION REFRIGERATION SYSTEM

A vapor compression refrigeration system is the most commonly used refrigeration system as it is feasible and readily available. It receives the evaporated refrigerant at temperatures below 0°C to obtain refrigeration. A VCRS system undergoes a phase change process of the refrigerant (working fluid) cycle of operation with four main components, which makes the cycle complete. The evaporator is used to generate a phase change of the low-pressure refrigerant to change its phase from liquid to vapor. This vapor is sent through a connecting pipe to the compressor, which is to pressurize the vapor and increase both temperature and pressure. The input power to the system is to run the compressor, which makes it a mechanical refrigeration system. The vapor is next passed through a condenser to reject heat from the working fluid. The condensed refrigerant changes its phase to a high-temperature/high-pressure liquid, and it is throttled or expanded to a high-temperature/low-pressure vapor by the throttling valve. The vapor compression refrigeration system is widely used for small and large-scale applications due to its refrigeration capacity and also its size and weight.

3.3.1 MAJOR COMPONENTS OF VCRS

The main components of a VCRS system are:

- Evaporator.
- Compressor.
- Condenser.
- Expansion device.
- Liquid receiver.
- Filter drier.
- Oil separator.

Evaporator

The evaporator is the energy producing (output) device that produces the refrigerating effect through the phase change of refrigerant. It will convert the low-pressure/low-temperature liquid into a low-temperature/low-pressure vapor. The heat from the substance or system is added to the refrigerant and increases the temperature. The higher the efficiency of the evaporator is, higher the refrigeration effect will be. There are different types of evaporators based on design and application:

- Bare tube evaporator
- Finned evaporator
- Plate-type evaporator

Compressor

The compressor is the energy absorbing device (input) that is actuated by a mechanical shaft powered by a motor. It draws the refrigerant from the evaporator and compresses it to a high-temperature/high-pressure vapor. The lower the compressor work is to compress the refrigerant, the higher the efficiency of the VCRS system will be. The compressor capacity plays a vital role in the selection of refrigeration system.

Compressors are classified as follows:

BASED ON THE WORKING PRINCIPLE

- Positive displacement type
  - Reciprocating type
  - Rotary screw type (twin screw or single screw)
  - Rotary type with sliding vanes (multiple or single vane, rolling piston)
  - Orbital
  - Acoustic
- Rotodynamic compressor
  - Radial flow
    - Centrifugal compressor
  - Axial flow

BASED ON THE ARRANGEMENT OF MOTOR DRIVE

- Open type
- Hermetic (sealed)
- Semihermetic (semisealed)

Open-type Compressors

These compressors are connected externally through a shaft to a motor drive, driven by a belt or gear. These compressors are large as they are used for medium- to large-scale application. Since it is an open type, it is easy to service and maintain these compressors. These provide good efficiency. As the shaft is projected out from the compressor, there is some inevitable loss of refrigerant during operation. Hence there is a need for a refrigerant reservoir.

Hermetic Compressors

In this type of compressor, the motor drive and compressor are enclosed in a vessel that is completely sealed to avoid leakage. In the hermetic compressor, motor heat is rejected and cooled by the gases in the compressor. The refrigerant flows over the winding of the motor so that not all refrigerants can be used in a hermetic compressor. Only a refrigerant with dielectric properties is allowed to be used in these compressors. These are maintenance-free and nonserviceable compressors that can be used only for a range of temperature operations. It is used in small-scale and residential refrigerators. The efficiency of the hermetic compressor is lower than that of the open type, but it is cost-efficient.

Semihermetic Compressors

These are same as the hermetic type with a removable lid or cylinder head so that the piston and motors can be serviced easily. These compressors work at constant speeds, and no speed control is possible.

Condenser

The condenser is the heat exchanging device used to exchange or remove heat from the refrigerant with water or air as the cooling medium based on the application to ensure high rate of heat removal from the refrigerant. It converts the high-temperature/high-pressure vapor to a high-temperature/high-pressure liquid by enhancing phase change. The heat removed from the refrigerant by the condenser is rejected to the atmosphere. Condensers are classified, based the condensing fluid, as follows:

- Evaporative condenser
- Water-cooled condenser
  - Tube in tube type
  - Shell and tube type
  - Shel and coil type
- Air-cooled condenser
  - Natural convection type
  - Forced convection type

Expansion Device

Expansion is the process of reducing the pressure from the liquid refrigerant that exits from the condenser and it is sent to the evaporator. The high-pressure/high-temperature liquid is expanded to a low-pressure/low-temperature liquid by sensible heat removal. In reality, the expanding device will also handle both liquid and gaseous refrigerants by changing their areas and mass flow rates.

Based on the refrigeration system, expansion devices are classified as follows:

- Fixed opening type
  - Capillary tube
  - Orifice
- Variable opening type
  - Thermostatic expansion valve (TEV)
  - Electronic expansion valve
  - Automatic expansion valve (AEV)
  - Float-type expansion valve
    - High side float valve
    - Low side float valve

Oil Separator

Oil separators are installed in the compressors to remove the accumulation of oil from the refrigerant at the outlet of the compressor. It intercepts the oil from the refrigerant and returns it to the crankcase of the compressor to assure the lubrication of all components.

Liquid Receiver

The liquid receiver is the reservoir to store the liquid refrigerant coming out of the condenser. It will transfer the refrigerant to the expansion device through a metered flow control device. The liquid receiver is situated lower than the condenser in order to accumulate the drained refrigerant. It will be available with the pressure relief valve to avoid blockage.

Filter Drier

he filter drier is a moisture drier setup at the coolest area on the system to remove moisture dirt from the refrigerant before entering the thermostatic expansion valve. The filter drier will be between the condenser and the expansion device.

Filter driers are classified as follows:

- Liquid line type
  - Straight-through sealed type
  - Replaceable core type
- Suction line type

3.3.2 PROCESS

The vapor compression refrigeration system follows the reverse Rankine cycle.

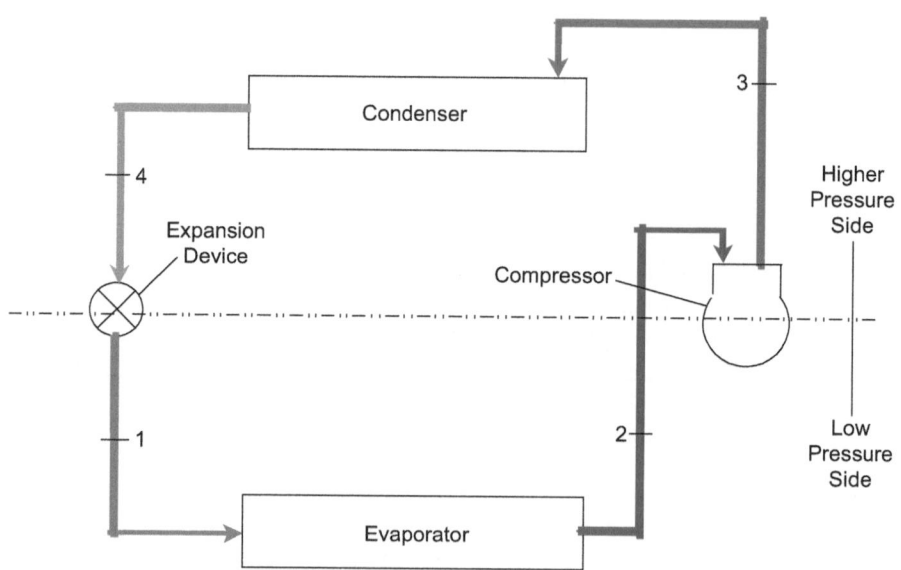

FIGURE 3.1 Vapor compression refrigeration system process.

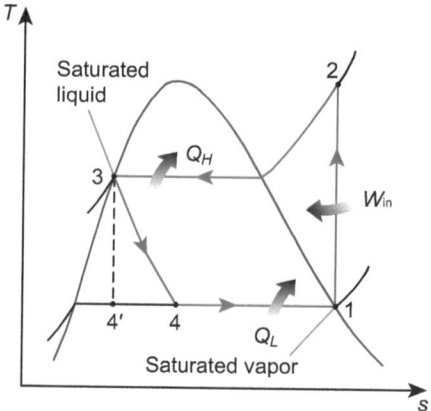

FIGURE 3.2 Standard vapor compression refrigeration system.

Process 1–2: Isentropic Compression

Compressor work done at saturated condition with constant entropy. The refrigerant at the low-pressure/low-temperature dry saturated vapor is compressed to a high-pressure/high temperature superheated vapor.

Process 2–3: Constant Pressure Heat Rejection

The high-pressure high temperature superheated vapor is then cooled or condensed to a saturated vapor, as shown in Figure 3.2, where sensible heat is rejected to the ambience, and further it is condensed to a high-pressure and high-temperature saturated liquid shown in the saturation curve.

Process 3–4: Isentropic Expansion

The saturated liquid refrigerant is further cooled to a subcooled temperature and stored in a liquid receiver. The low-temperature/high-pressure liquid refrigerant is sent to an expansion device such as thermostatic expansion valve. The latent heat of the refrigerant is rejected in (process 3–4), and further sensible heat is removed by reducing the pressure by altering the flow rate of the refrigerant as low pressure and temperature.

Process 4–1: Constant Pressure Heat Addition

The main process of the refrigeration cycle is the amount of heat extracted by the refrigerant from the refrigerated space. The refrigerant, a low-temperature and low-pressure liquid, exchanges heat with the evaporator, making it a low-pressure and low-temperature vapor. The increase in evaporator temperature increases the refrigerating effect.

3.3.3 Refrigerants

Refrigerant is a fluid or mixture of two or more fluids that undergoes phase change easily and that is used in refrigerators and heat pumps. They are classified into primary and secondary refrigerants.

Primary refrigerants are those used as working fluids in direct contact with refrigerated space. They undergo phase change and are used for varying temperature conditions. These are direct expansion systems. The classification of primary refrigerants is as follows:

- Halo-carbon refrigerant
- Azeotrope refrigerant
- Inorganic refrigerant
- Zoetrope refrigerant
- Hydrocarbon refrigerant

Halo-carbon Refrigerant

These are synthetically manufactured and developed freon refrigerants. They use the methane and ethane series of fluorocarbons. They contain one or more halogen atoms (chlorine, bromine, and fluorine. They are nonflammable, nonirritant to humans, and also nonexplosive in nature.

Chlorofluorocarbons: R11—trichloro-monofluoromethane
Hydrochlorofluorocarbons: R22—monochloro-difluoromethane
Hydrofluorocarbons: R134A—tetrafluoro-ethane

Inorganic Refrigerant

These are the oldest form of refrigerants, used before the development of halocarbon. These are still produced and used for their thermophysical properties. They are cheap and robust. These refrigerants have less ozone depletion potential.

Water
Air
Ammonia
Sulfur dioxide

Azeotrope Refrigerant

These are a group of two or more refrigerants whose pressure and temperature are similar and that have their fixed thermodynamic properties. A mixture of two or more refrigerants with liquid and vapor phases retain identical compositions over a range of temperatures.

Azeotropic refrigerants are designated as the R-500 series:

R-500: (73.8% R12 and 26.2% R152)
R-502: (8.8%R22 and 51.2% R115)
R-503: (40.1% R23 and 59.9% R13)

Zeotropic Refrigerant

Zeotropic refrigerant is a mixture in which the composition of one liquid phase differs from that of one vapor phase. They do not boil at a constant temperature.

Zeotropic refrigerant is designated as the R400 series:

R404A: R125/R134A/R143A (44%, 52%, 4%)
R407C: R134A/R125/R32 (23%, 25%, 52%)
R410A: R32/R125 (50%, 50%)
R413A: R600A/R218/R134A (3%, 9%, 88%)

Hydrocarbon Refrigerants

These are the most widely used and readily available refrigerants. These are used in industries and commercial applications. These refrigerants are highly explosive and flammable. They possess good thermodynamic properties. Generally, they are used in automobile air conditioning systems.

R170: Ethane (C_2H_6)
R290: Propane (C_3H_3)
R600: Butane (C_4H_{10})
R600A: Isobutane (C_4H_{10})

Secondary refrigerants transfer heat from the primary refrigerants. Generally, brines and water are used as secondary refrigerants because of operating temperature. They cool the substance by absorbing sensible heat and indirectly exchange heat. Secondary refrigerants are used in indirect expansion systems.
Calcium chloride
Sodium chloride

Requirements of Refrigerants

- Leak detection
- High performance
- Nonflammability
- Cost and availability

Refrigerant Nomenclature

All refrigerants are designated by **R** followed by a number: **R XYZ**, where:

$$X^{+1} = \text{Number of carbon atoms,}$$
$$Y^{-1} = \text{Number of hydrogen atoms, and}$$
$$Z = \text{Number of fluorine atoms.}$$

Example: **R22 — Monochlorodifluoro methane**

$$X = 0, Y = 2, Z = 2$$
Number of carbon atoms $= 0 + 1 = 1$
Number of hydrogen atoms $= 2 - 1 = 1$
Number of fluorine atoms $= 2$

The chlorine balance: **4-Number of (Hydrogen + Fluorine) atoms** = 4–1–2=1
The chemical formula: R22 = $CHCLF_2$

3.4 ENERGY CONSERVATION OPPORTUNITIES AND CASE STUDIES

3.4.1 Replace Belt-Driven AHU Drives with Direct Drives

Air handling units are an important part of the HVAC system. These are used to maintain appropriate conditions (relative humidity and temperature) in the building space and server room. The AHU fan is the major component that cools the return air from the building space.

Observations

AHUs have V-belt-driven fans. In any motor-driven equipment, inherent transmission losses are about 5% in the V-belt system. The details of the AHUs with the belt-driven drive are shown in the following table:

S. Number	Location	Running KW
1	AHU-1	7.65
2	AHU-2	1.2
3	AHU-3	5.78
4	AHU-4	4.27
5	AHU-5	5.75

Advantages of EC Direct Driven AHUs

- Motor efficiency full load > 95%
- Power factor at drive level close to unity
- Electronically commutated motor
- Backward curved fan blades

Proposed Solution

Replace the V-belt-driven fans with a direct driven type, as transmission losses are less in a direct driven system.

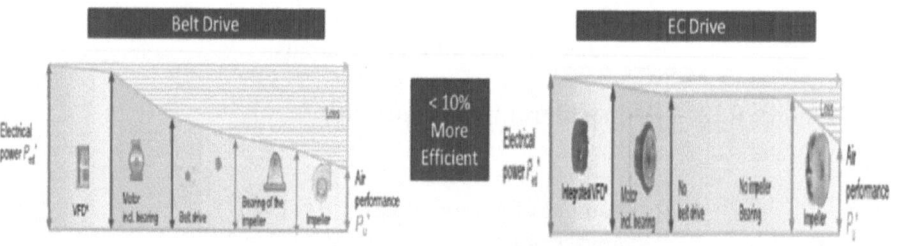

FIGURE 3.3   Replacement of belt-driven AHUs with direct drive.

Option 1: It is recommended to replace the fan shaft and directly couple it with the motor shaft. This will eliminate the transmission losses associated with the V-belt-driven system.

Option 2: It is recommended to replace the existing fan with a new PM axial fan, which has higher efficiency compared to all other types of fans due to its better aerodynamic design.

Cost Economics

Annual cost savings	Rs. 3.45 lakhs
Investment	Rs. 8.00 lakhs
Payback period	<2 years

3.4.2 ACTIVE REFRIGERANT AGENT TO AVOID OIL FOULING

Observations

At present, there exists VRFs (variable refrigerant flows) and split/window ACs apart from chillers to cater to the cooling load of the entire engineering facility, which is more than 6 years old.

Axial Fans
▪ Vane axial fans

FIGURE 3.4 Direct drive fans.

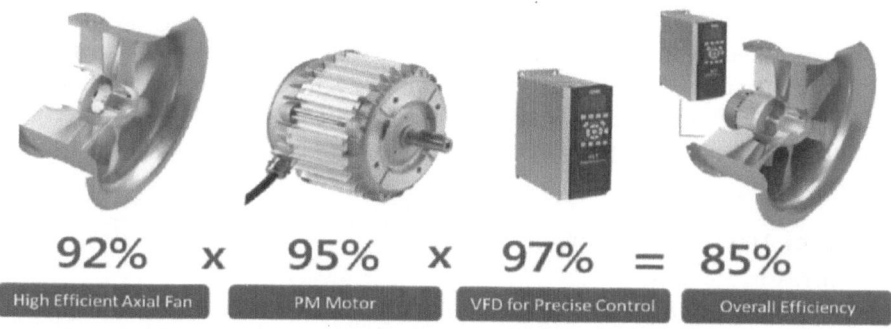

92% x 95% x 97% = 85%

| High Efficient Axial Fan | PM Motor | VFD for Precise Control | Overall Efficiency |

FIGURE 3.5 Energy efficiency of direct drive fan unit.

Performance

Sl. Number	Particulars	Power (kW)
1	VRF-1	6.26
2	VRF-2	13
3	VRF-3	5
4	VRF-4	8.76
5	VRF-5	15.26
6	VRF-6	4
7	Split/window ACs	62.4

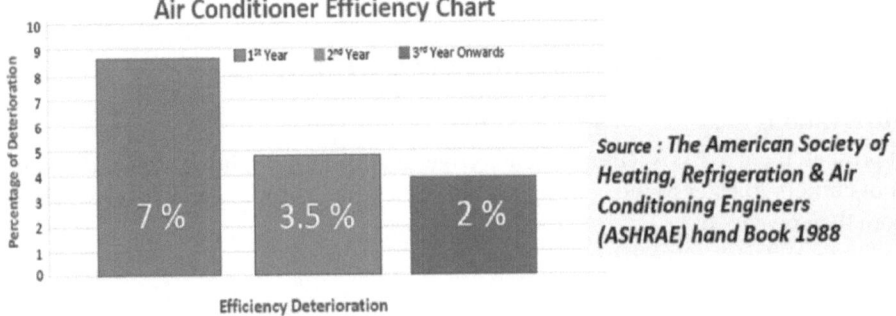

FIGURE 3.6   Air conditioner efficiency chart.

Oil Fouling

During the piston stroke in the cylinder of a compressor, the refrigerant gets a small amount of oil from the compressor, which, over time, adheres to the inner pipe of the evaporator. That adhesion oil acts as an insulator, reducing the heat transfer capacity of the evaporator.

Typically, an AC's efficiency deterioration is more than 30% for a 20-year-old system.

Active Refrigerant Agent

This is an intermetallic compound technology that, when introduced into the refrigerant oil, forms a permanent bond to metal surfaces. It removes oil fouling, changes the thermal nature of the metal, and lowers the boiling point of the refrigerant gas, resulting in a more efficient operating system with substantial energy cost savings.

Benefits of Active Refrigerant

- Increases lubricity
- Reduces friction between mechanical parts, thus increasing their life and efficiency
- Reduces wear and tear

- Low maintenance
- Treats the metal surfaces and sends debris back to the filtration system
- Equipment damage prevention during a catastrophic loss of primary lubricant
- Withstands pressure of 4500 psi
- Provides continuous lubrication

Proposed Solution

We recommend adding an active refrigerant agent for the chiller that has the potential to save around 6% of energy consumption.

Cost Economics

Annual cost savings	Rs. 11.40 lakhs
Investment	Rs. 3.60 lakhs
Payback period	>1 year

3.4.3 INSTALL VFD FOR PRIMARY CHILLED WATER PUMP

Observations

A schematic of the chilled water system is shown in Figure 3.8. Primary chilled water pumps deliver chilled water from the AHUs into the chillers. The secondary chilled water pumps deliver water from the chillers to the AHUs. The decoupler is an essential interconnection line that bypasses the users when required and avoids icing inside the chilled water pipeline by maintaining minimum flow across the chiller.

The system has a VFD (variable frequency drive) for the secondary chilled water pump. The primary chilled water pump is not installed with VFD, and a flow of 85 m³/hr is maintained constantly in the system. The secondary chilled water pump

Before **After**

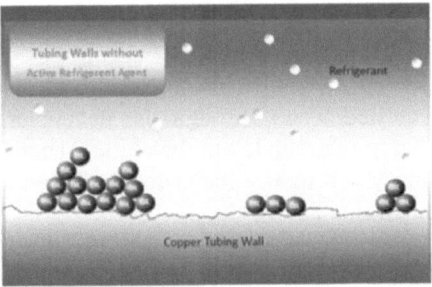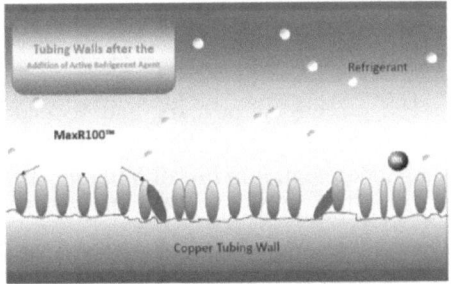

FIGURE 3.7 Effect of active refrigerant agent (before and after).

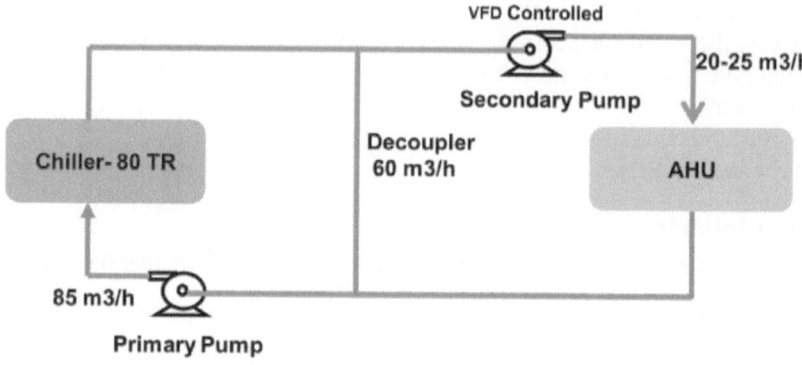

FIGURE 3.8 Primary chiller water pump process.

FIGURE 3.9 Variable frequency drive (VFD).

delivers a flow of 25 m³/hr to the user. The excess flow is through the decoupler and is around 60 m³/hr.

Chiller	Primary chilled water flow (m³/hr)	Secondary flow (m³/hr)	Flow through decoupler (m³/hr)
Chiller 1	85	25	60

The mixing of primary and secondary chilled water through the decoupler reduces the temperature of the chilled water being recirculated in the chiller. There is a good potential to install VFD for the primary and reduce the decoupler flow to less than 10% of the total flow delivered by the primary.

Proposed System

The primary and secondary flows can be maintained at the same rate. The ΔT of the primary and secondary sides can also be maintained. The reduction of flow through the decoupler will reduce the power consumed by the primary pump by 20%. The additional advantage of installing VFD for the primary is that it will reduce the mixing of temperatures through the decoupler and improve the operation of the chiller. Care should be taken that the minimum flow through the chiller is always maintained and that VFD should interlock and not allow the VFD to go below minimum flow through the chiller.

Proposed Solution

Equalize the flow between the primary and secondary chilled water pumps and minimize the flow through the decoupler, maintaining the flow at less than 10%.

Cost Economics

Annual cost savings	Rs. 1.00 lakhs
Investment	Rs. 0.40 lakhs
Payback period	>1 year

3.4.4 INSTALL VFD FOR CHILLER COMPRESSOR

Observations

Two 80 TR (tons of refrigeration) chillers are installed to meet the space cooling requirements in the engineering facility. One of the chillers remains on standby for the ad hoc requirement. The chiller is an air-cooled screw compressor operating on a load–unload condition and has no control for precise capacity control as per requirements. At present, the chiller is switching off intermittently, indicating a higher-capacity chiller against the requirement.

The details of the chiller are shown in the following table:

Chiller number	Type	Rating
1	Air-cooled screw	80 T
2	Air-cooled screw	80 T

The chiller has an on/off cycle that can increase the mechanical wear and tear of the moving parts and thereby reduce the life cycle of the equipment. The power consumption also increases due to inrush current. There is no VFD installed in the compressor, and thus load–unload is taking place during low demand conditions.

Proposed Solutions

Install VFD in the chiller compressor to precisely match the load requirements. The compressor unloading will be avoided, and the speed of the compressor will be adjusted as per the load requirement.

FIGURE 3.10   Variable frequency drive (VFD).

Cost Economics

Annual cost savings	Rs. 4.50 lakhs
Investment	Rs. 3.60 lakhs
Payback period	>1 year

3.4.5 INSTALL ADIABATIC COOLING SYSTEM FOR CHILLERS

Observations

The chiller system at the engineering facility consists of two 80 TR chillers, which are air cooled. One chiller is in running condition, and the other is in standby condition. The chiller installed has been operating with a specific energy consumption of 1.28 kW/TR, which is high for an air-cooled chiller.

The specific energy consumption of a water-cooled chiller is around 0.6 kW/TR. A water-cooled chiller system usually consumes less than 0.7 kW/TR, due to the lower temperature of the medium of heat transfer (water). There is a good potential to achieve efficiency closer to water-cooled chiller by installing an adiabatic cooling system.

Proposed System

An adiabatic cooling system has been installed by the plant. Due to water contamination, the system has not been in use. The schematic of the adiabatic cooling system is shown in Figure 3.11. The adiabatic system consists of fine water spraying in a mesh around the air-cooled condenser. Due to this water spray and mesh, the entering temperature of the air around the condenser of the air-cooled condenser chiller is reduced due to evaporative cooling.

As the air temperature to the condenser is reduced, the performance of the chiller improves by lowering the discharge pressure of the compressor, as the refrigerant

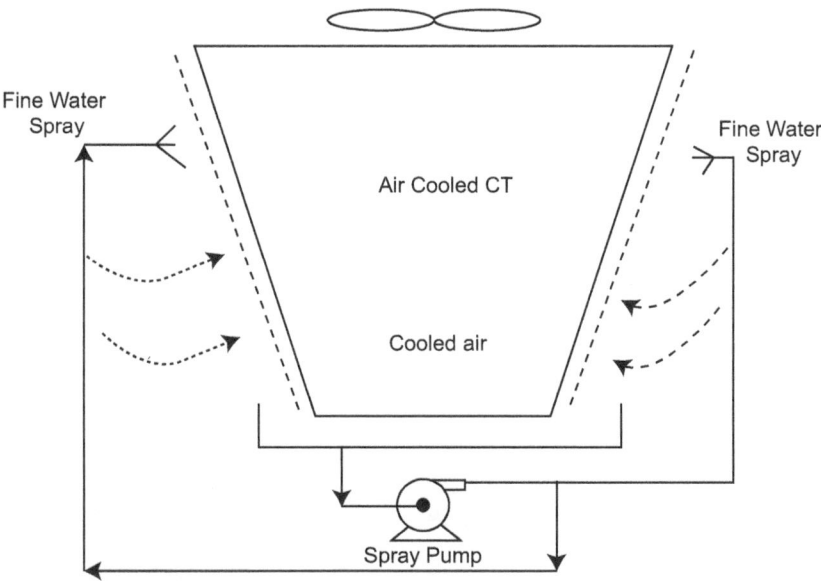

FIGURE 3.11   Adiabatic cooling system process.

can be condensed at a lower saturation temperature, thus reducing the power consumption of the chiller. In the case of the air-cooled condenser with fogging, care
should be taken that sufficient air ventilation is available across the condenser for the
effective operation of the adiabatic system.

Installation of an adiabatic cooling system is shown in Figure 3.12. This adiabatic
system is recommended for installation because the air-cooled condenser receiving the cooled air close to wet bulb temperature is completely taken through the
adiabatic cooling pads, wherein the water is sprayed to reduce the air temperature
for condensation. Also, the opening on the sides of the air-cooled chiller is closed to
ensure that all the air is taken through the adiabatic cooling system.

Proposed Solution

Install an adiabatic condenser for air-cooled chillers to ensure that all air is passed
through the adiabatic cooling system. The installation of an adiabatic cooling system can
improve the performance of an air-cooled chiller close to that of the water-cooled chiller.
A 15% reduction in energy consumption of the chiller compressor can be obtained.

An additional installation of the softener system is recommended to improve the
quality of water that will be fed to the adiabatic cooling system.

Cost Economics

Annual cost savings	**Rs. 2.20 lakhs**
Investment	Rs. 3.50 lakhs
Payback period	>2 years

FIGURE 3.12   Adiabatic cooling system installation.

3.4.6 INSTALL AC ENERGY SAVERS FOR AC UNITS IN PLANT

Observations

Connected AC Details

TR rating	Numbers	Total TR
1	13	13
1.5	03	4.5
2	11	22
3	1	3
5	2	10
Total	30	52.5

Air Conditioner Energy Saver

Install an AC energy saver, which is a dual-sensor module for split ACs. It is a program-mable microprocessor-based AC energy saver. Equipped with dual sensors, this can read and display both room and coil temperatures, with reference to the ambient temperature.

The energy saver's dual sensing technology reads the room, coil, and ambient temperature. With its multiple algorithms in a closed-loop circuit, it ensures high savings and adapts your AC to ambient temperatures and climatic changes.

Conventionally packed AC units have the following problems:

- AC manufacturers cannot customize each unit to the different climates and therefore design a common control setting for the hottest conditions. This leads to huge wastage. When the set temperature is achieved, the compressor continues to run for an additional fixed period (6–8 minutes—known as the overcooling period), which is required for only a few hours in a day in peak summer but which is overutilized the balance of time. Also, in hotter climates, by reducing the prefixed off-time, the overcooling can be reduced substantially.
- Many ACs typically do not achieve the set temperature, especially if it is set at 18°C or 19°C. As a result, the compressor runs continuously, resulting in wastage of a huge amount of electricity, giving rise to issues such as:
  - Ice formation on the coil and reduced heat transfer and cooling.
  - Motor and compressor running at higher temperatures, increasing the specific energy consumption by 30–40%.
  - Refrigerant liquefying and leading to the risk of compressor seizing.
  - Frequent maintenance issues

Energy saver lets you program your off-time and overcooling period based on your climate and day/night usage. The coil sensor in the energy saver will cut off the compressor at a coil temperature programmed by the user and not only saves on energy consumption but also increases the life of the compressor and reduces the number of breakdowns.

Proposed Solution

Install an AC energy saver for identified AC units in the plant in a phased manner, and optimize the power consumption of existing air conditioning units.

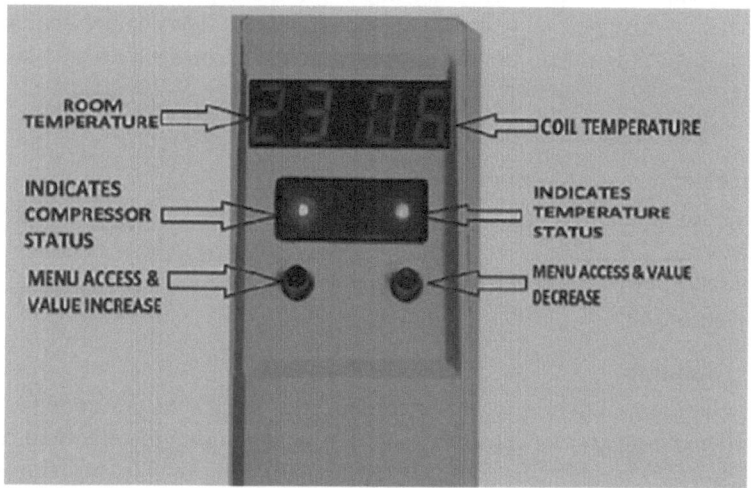

FIGURE 3.13 Air conditioner energy saver.

Cost Economics

Annual cost savings	Rs. 4.50 lakhs
Investment	Rs. 4.40 lakhs
Payback period	>1 year

3.4.7 INSTALL SOLAR THERMAL HYBRID FOR VRF SYSTEM

Observations

The major VRF units are installed for different room for meeting variable cooling requirements. The details of the power consumption of the VRF units are shown in the following table:

Area	Present power (kW)
VRF-1	6.26
VRF-2	13
VRF-3	5
VRF-4	8.76
VRF-5	15.26
VRF-6	4
Total	52

Solar Thermal Hybrid

The system is a renewable energy method of adding pressure and heat to the refrigeration cycle, which results in a decreased/displaced compressor workload, saving energy. This solar thermal system displaces a portion of the mechanical energy used by various compressor types, including single-speed compressors with VFDs, variable capacity, multistage, and variable speed compressors. The compressor can operate at low stage, low range, or low capacity, while delivering full- and part-load cooling requirements, creating significant energy savings of 25–40% per year, or more.

Advantages

- A unique blend of superior energy efficiency with a renewable energy platform
- Upgrades for new and existing cooling systems
- Energy savings of 25–40%

Proposed Solution

Install a solar thermal hybrid for the VRF units, and reduce the power consumption of the chiller compressor by extracting some work done using the renewable energy medium.

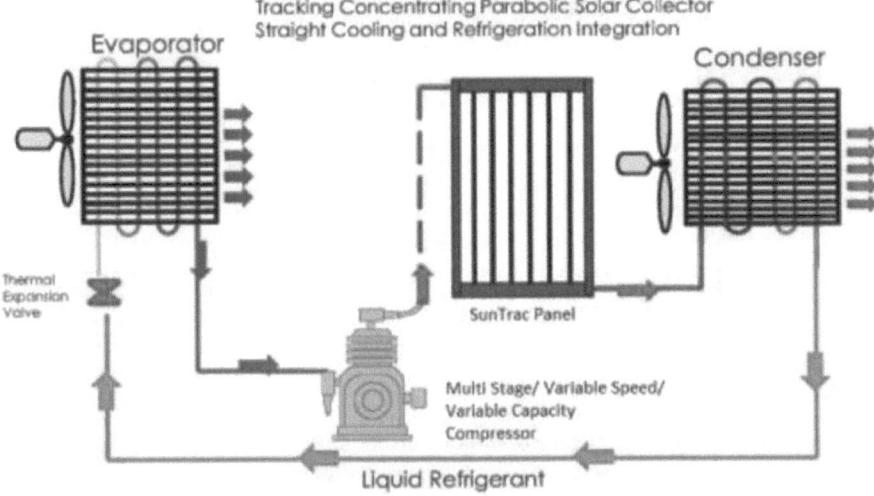

FIGURE 3.14 Solar thermal hybrid for VRF systems.

FIGURE 3.15 Solar thermal hybrid system.

Cost Economics

Annual cost savings	**Rs. 4.60 lakhs**
Investment	Rs. 14 lakhs
Payback period	<2 years

3.4.8 Install Adiabatic Cooling System for VRF Out Door Units

Observations

Area	Present power (kW)
VRF-1	6.26
VRF-2	13
VRF-3	5
VRF-4	8.76
VRF-5	15.26
VRF-6	4
Total	52

The specific energy consumption of a water-cooled chiller is around 0.6 kW/TR. A water-cooled chiller system usually consumes less than 0.7 kW/TR due to the lower temperature of the medium of heat transfer (water). There is a good potential to achieve efficiency closer to that of a water-cooled chiller by installing an adiabatic cooling system.

Proposed System

An adiabatic cooling system has been installed by the plant. Due to water contamination, the system has not been in use. The schematic of the adiabatic cooling system is shown in Figure 3.16. The adiabatic system consists of fine water spraying in a

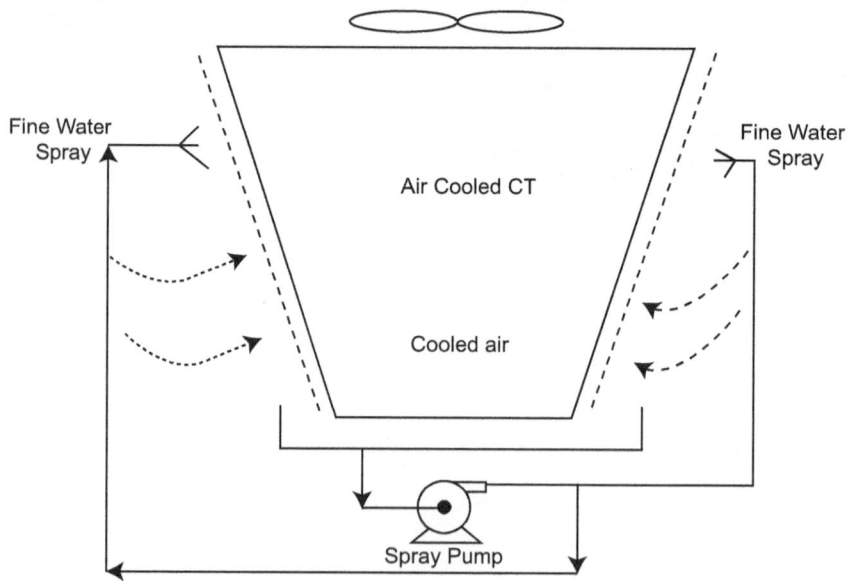

FIGURE 3.16   Adiabatic cooling system for VRF ODUs.

mesh around the air-cooled condenser. Due to this water spray and mesh, the entering temperature of the air around the condenser of the air-cooled condenser chiller is reduced due to evaporative cooling.

As the air temperature to the condenser is reduced, the performance of the chiller improves by lowering the discharge pressure of the compressor because the refrigerant can be condensed at a lower saturation temperature, thus reducing the power consumption of the chiller. In the case of the air-cooled condenser with fogging, care should be taken that sufficient air ventilation is available across the condenser for the effective operation of the adiabatic system.

Installation of the adiabatic cooling system is shown in Figure 3.17. This adiabatic system is recommended for installation because the air-cooled condenser receiving the cooled air close to wet bulb temperature is completely taken through the adiabatic cooling pads, wherein the water is sprayed to reduce the air temperature for condensation. Also, the openings on the sides of the air-cooled chiller are closed to ensure that the complete air is taken through the adiabatic cooling system.

Proposed Solution

Install an adiabatic condenser for VRF ODUs, ensuring that all the air is passed the adiabatic cooling system. Because it is a mini version of an air-cooled chiller, the adiabatic cooling system can improve the performance of VRF ODUs close to that of a water-cooled chiller. A 15% reduction in energy consumption of the chiller compressor can be obtained.

Cost Economics

Annual cost savings	**Rs. 2.30 lakhs**
Investment	Rs. 3.50 lakhs
Payback period	>2 years

FIGURE 3.17   Adiabatic cooling system.

3.4.9 INSTALL VARIABLE AIR VOLUME (VAV) SYSTEM FOR AHUs

Observations

The present power consumption of major AHUs are shown in the following chart:

S. Number	Location	Running kW
1	AHU-1	7.65
2	AHU-2	1.2
3	AHU-3	5.78
4	AHU-4	4.27
5	AHU-5	5.75
	Total	24.65

At present, the AHUs have a fixed damper control for the supply air to the space cooling requirement in the engineering facility. The AHUs are belt driven, and the cascaded efficiency is low due to transmission losses. Fixed dampers do not provide control of airflow based on occupancy demand requirements.

Advantages of VAV

- Motorized valves
- Option to include various external sensors as feedback to the valves
- Better management of the temperature gradient in the system
- Remote monitoring options—air conditioning management system

Proposed Solution

Install a VAV system for the major AHU units in the engineering facility. The VAV will help in precise airflow management based on temperature and occupancy level in the rooms, thereby improving the efficiency of the system.

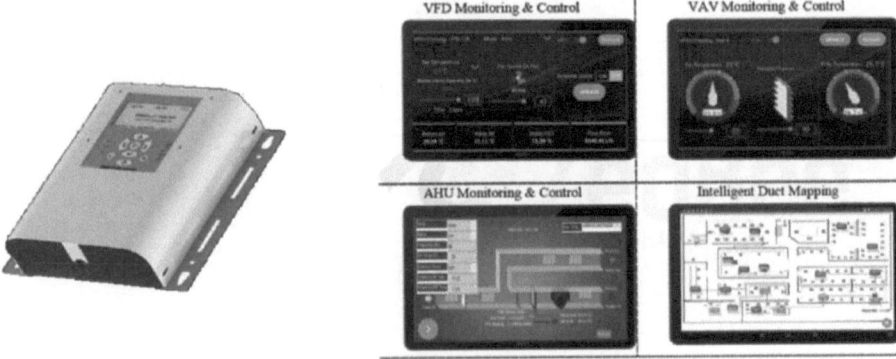

FIGURE 3.18   VAV system for AHUs.

Cost Economics

Annual cost savings	Rs. 1.80 lakhs
Investment	Rs. 3.00 lakhs
Payback period	>2 years

3.4.10 INSTALL HYDROMX SYSTEM FOR CHILLERS

At present there exists three 370 TR air- and water-cooled chillers, which were of more than 8 years old, to serve the cooling load of the entire facility. The units consume around 350 units per hour.

There is a scope for reducing operating energy cost by increasing the heat transfer rate in the chilled water system.

Hydromx

Hydromx is a heat transfer fluid that provides significant energy savings in heating and cooling through its nanothermo technology. Hydromx uses nanoparticles that are suspended in a stable state to increase the speed of heat transfer by heating up (or removing heat from) the fluid and transferring energy in less time, thereby requiring significantly less energy (and saving money).

Hydromx is most suitable for closed-loop heating and cooling systems where efficiency is achieved by diminishing the run times of associated equipment. Hence the life of the equipment is extended, and maintenance costs are lowered.

Furthermore, Hydromx is specifically formulated to prevent corrosion, calcification, and algae in the systems. It is certified under the BuildCert Chemical Inhibitor Approval Scheme to inhibit corrosion of metallic and plastic parts and, particularly, to prevent scaling in the system.

Hydromx in Cooling Systems

In cooling systems, when compared with water, Hydromx makes heat transfer 35% faster, transferring the same amount of heat in a much shorter time. The energy efficiency of Hydromx is based on reducing the run time of the compressor, and Hydromx increases the required ambient temperature limits for free cooling up to 5–6°C. It is possible to achieve up to a 35% saving in cooling with Hydromx and as high as 37% in highly controlled environments such as a facility.

Features of Hydromx
- Operating range of −60°F to +244°F
- Reduced carbon emission
- 20–35% in energy savings
- Increased equipment life
- Rapid return on investment

Proposed Solution

Install Hydromx for the chiller, which has the potential to save energy consumption by optimizing the compressor running cycle for the same cooling load.

Cost Economics

Annual cost savings	Rs. 15.00 lakhs
Investment	Rs. 45.00 lakhs
Payback period	<2 years

3.4.11 Install Free Cooling System

Chilled water is supplied to a PAHU (precision air handling unit), which maintains the required cold aisle temperature of 24°C and humidity within the range as per ASHRAE standards. Currently, the entire facility cooling is served by the chiller and PAHU system, consuming nearly 307 KW of instantaneous power. Chiller and PAHU units run almost 365 days, 24 hours a day at different load settings.

Average hot aisle temperature at the time of the audit was found nearly 32°C. After analyzing the weather data of geographic location, we found a scope for employing a free cooling system and thus avoid full load on chilled water system for some hours throughout a year.

- Minimum temperature: 15°C
- Mean temperature: 27°C
- Humidity: 52%

Proposed Free Cooling System

In this proposed system, a separate duct is joined to the existing duct, which will take ambient fresh air in to the facility along with air from chilled water. An exhaust system also needs to be provided to continuously replace the air inside. Due to favorable ambient conditions (lower temperatures than the hot aisle temperature) for at least 4 months in a year, this will reduce the cooling load on the chiller. Feedback systems should be provided to control the system in such a way as to attain required ASHRAE conditions inside.

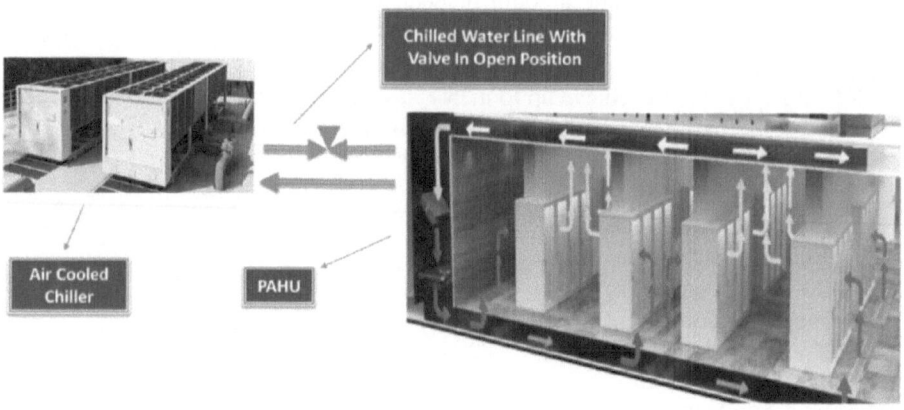

FIGURE 3.19 Chilled water supply in PAHU.

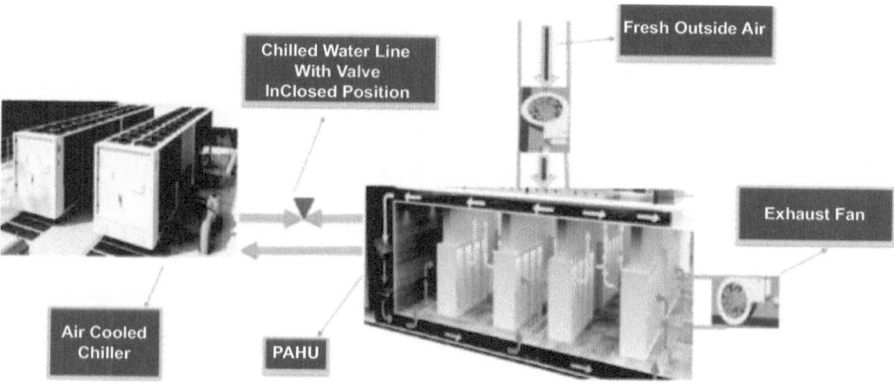

FIGURE 3.20 Proposed chilled water supply for PAHU.

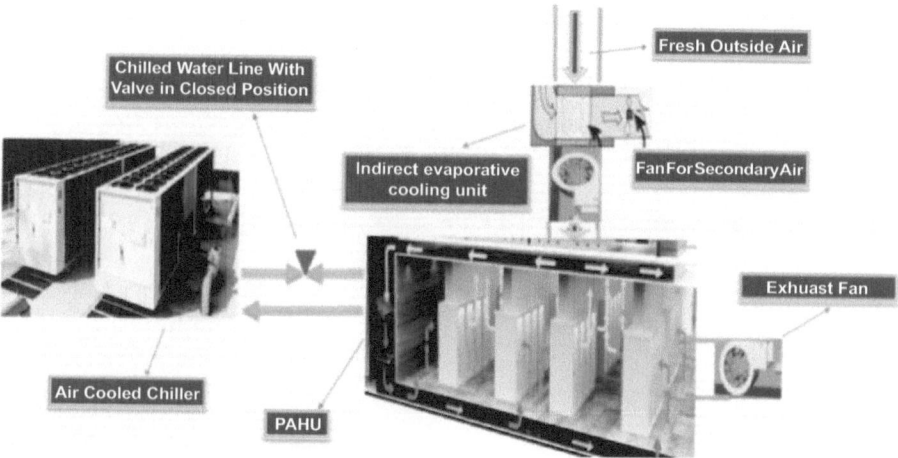

FIGURE 3.21 Indirect evaporative cooling.

For further enhancing this proposed system, the inlet fresh air can be precooled through indirect evaporative cooling and sent to the facility, which has the effect of a further load reduction on the chiller.

Indirect Evaporative Cooling

This is a kind of cooling system wherein the fresh ambient air is passed through a cooling coil to further reduce the temperature. And the cooling coil is supplied with water that will be circulated through the cooling tower.

Proposed Solutions

Install a free cooling system so that the facility can avoid running the chiller and PAHU unit for some hours in a year when the outside ambient temperature is less than the internal white space temperature.

Cost Economics

Annual cost savings	Rs. 25.00 lakhs
Investment	Rs. 25.00 lakhs
Payback period	1 year

4 Energy Efficiency in Furnaces

Sivakumar

CONTENTS

DOI: 10.1201/9781003203810-4

LEARNING OUTCOMES

At the end of this chapter, the reader will be able to understand:

- The basics about furnaces and their types.
- Energy conservation opportunities in furnaces.
- Energy monitoring and data analysis.
- Real-time case studies with saving calculations.

4.1 INTRODUCTION

A furnace is a device in which heat is generated and transferred to materials with the object of bringing about physical and chemical changes. Furnaces generate heat in a controlled manner through the combustion of a fuel source. The source of heat is usually combustion of solid, liquid, or gaseous fuel or electrical energy applied through resistance heating (Joule heating) or inductive heating. The thermal energy is then used to heat spaces such as rooms, buildings, or other structures. Other furnaces may be used in commercial and industrial settings for processing materials. In this chapter, the different types of furnaces are explored and summarized under two broad classifications:

1. Furnaces for heating homes and buildings
2. Industrial furnaces for metals and materials processing

4.2 INDUSTRIAL FURNACES

Furnaces in industrial settings heat up materials using fuel and combustion gases. The material may be in direct contact with the fuel and its gases (blast furnaces), in indirect contact with the fuel but still in direct contact with the gases (reverberator), or in indirect contact with both fuel and gases (muffle furnaces). Nevertheless, the goal remains the same—attaining a high level of heat.

One of the primary considerations in selecting an industrial furnace is typically the range of operating temperatures the furnace can provide. Furnaces reaching higher temperatures typically cost more. What you are looking for is a furnace that can efficiently and uniformly achieve the temperatures you need.

Some of the most common types of furnaces used for metals and materials processing are:

- Bell.
- Box.
- Forging.
- Pit.
- Quenching.
- Rotary.

- Salt bath.
- Tempering.
- Vacuum.

4.2.1 Bell Furnaces

Bell furnaces are electrically heated, gas-fired, or thermal recirculation heating systems featuring a movable dome. This furnace hood can be lifted frequently throughout vacuum or controlled atmosphere processes. Bell furnaces are used to temper, anneal, normalize, and stress-relieve steel plate formed parts. This equipment can be used with multiple bases and still provide reliable seals. These furnaces (multi- or single stack) are used to treat coils, strips, sheets, rods, and more.

4.2.2 Box Furnaces

Used for heat-treating, calcining, curing, preheating, tempering, and other processes, box furnaces feature vertical-lift or swing-open doors insulating the furnace and ensuring consistent airflow management. Box furnaces, which accept heavy loads, are set up to be loaded by forklift, roller hearth, or other manual processes. Gas and electric versions are available. Able to uniformly treat large volumes of material, box furnaces are seen in laboratory and industrial applications.

4.2.3 Forging Furnaces

Often used for preheating, slot forging furnaces resemble a pizza oven. These open-hearth furnaces transmit heat via radiation to get metals to a temperature at which they can be forged or hardening processes can be stopped. Heavy-duty forging furnaces heat and reheat large steel ingots, blooms, and parts. These furnaces can also come in forging box style. Load charge and discharge can be automated in some of these furnaces.

4.2.4 Pit Furnaces

Available in gas-fired or electric-powered configurations, pit furnaces are also known as top load furnaces. Offered in many sizes, pit furnaces heat to different temperatures within a secured working chamber with a controlled atmosphere. These furnaces are often used in automotive and aircraft part manufacturing, as well as in machine building, wind energy, and mining.

4.2.5 Quenching Furnaces

Quenching furnaces feature an enclosed heating chamber to prevent low-temperature processes, such as phase transformations. The furnace's controlled, rapid cooling hardens the material. This furnace process aims to avoid uneven heating and overheating, but the tempering technique may be performed after quenching to increase toughness. Products processed in quenching furnaces can include gears, bearing components, fasteners, construction, and agricultural machinery components. Quenching furnaces are available in electric or gas-fired models.

4.2.6 ROTARY FURNACES

Rotary hearth furnaces have a lined furnace generally welded from steel into a barrel shape. The furnace is mounted on a drive that rotates the barrel throughout the heat treatment. The material sample can be mixed by tilting the furnace during rotation. Highly heat efficient with easy pressurization, these furnaces provide heat uniformity and good material contact. The internal heat source can be gas or electric, with the flow of muffled combustion gases often countercurrent. Applications typically include calcination and oxidation.

4.2.7 SALT BATH FURNACES

Salt bath furnaces use the high-heat transfer characteristics of convection to achieve very fast heating of metal parts in heat treating applications. These furnaces are almost invariably electric as electrodes can be immersed directly in the molten salt, although externally heated electric or gas-fired units are sometimes used. The temperature of the bath depends on the varieties of salts used, which are commonly cyanide mixtures and chloride mixtures, with possible temperatures of 300–2350°F. Salt bath furnaces are used for treating high-speed tool steel and other edge tools, annealing nickel-chrome alloys and stainless steel, austempering and martempering steel, etc., as well as for brazing of otherwise difficult to braze parts such as automotive radiators or for setting the shape of shape-memory alloys.

4.2.8 TEMPERING FURNACES

Designed to heat-treat ferrous metal products, tempering furnaces increase toughness. Heat-treating certain high-strength materials can impact the alloy's formation and energy absorption.

To access the best balance of strength and elasticity, tempering furnaces are often used in conjunction with quenching furnaces. It is crucial that these furnaces maintain uniform temperature levels throughout the chamber to achieve the desired material characteristics. Both gas and electric heating types of these furnaces offer indirect fuel contact.

4.2.9 VACUUM FURNACES

Vacuum furnaces are used in many industries. What distinguishes this type of furnace is the vacuum maintained throughout the heating process to protect heated steel and metal parts. The furnace can be electric or gas heated, with pumps preserving the vacuum to prevent oxidation, heat loss, or contamination. These furnaces are used for annealing, brazing, sintering, and heat treatment.

4.2.10 OTHER TYPES OF INDUSTRIAL FURNACE

The many more various types of industrial furnaces include shell baking, laboratory, conveyor, pusher, raised or roller hearth, and industrial ovens and driers.

Industrial furnaces can also be categorized by application to include aluminizing, metal melting, brazing, calcination, among others.

4.3 CAST IRON MELTING AND ENERGY CONSERVATION

4.3.1 Cast Iron Melting Furnace

Figure 4.1 shows the kinds of cast iron melting furnace. Today, the actual percentage of cupolas and induction furnaces used for cast iron melting is obscure due to a lack of proper statistics. Systems with a cupola that were used for primary melting totaled about 85% in the 1970s, but their number was remarkably reduced to 63% around 1980. After that, cupolas may be used at a percentage of 50–60%.

The cupola is a shaft furnace for the continuous melting of cast iron with new pig iron, return scrap iron, and steel scrap used as raw materials and coke used as a fuel. A cupola has not only an economic advantage of low equipment cost but also refining and self-purifying capability, which makes it possible to get excellent molten metal even from inferior-quality raw materials, and it has therefore been widely used.

However, the exhaust gas from cupolas, containing not only carbon dioxide from coke combustion but also smoke- and dust-generated coke ash, NO, and SO, causes pollution in the air and the working environment. The use of a solid fuel makes it difficult to manage and control operation, making it difficult to cope with mechanization, personnel saving, and especially lack of skilled workers.

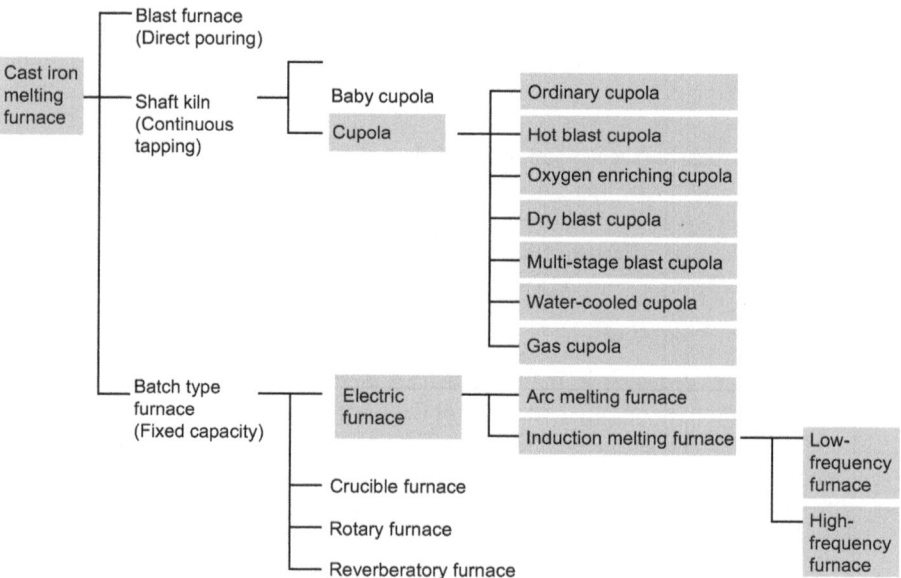

FIGURE 4.1   Classification of cast iron melting furnaces.

4.3.2 ENERGY CONSERVATION IN MELTING

Independently of the kind of melting furnace, the following considerations are necessary for energy conservation in the melting process:

1. Improvement of melting operation
2. Reduction of heat input
3. Reduction of heat loss

4.4 CUPOLA MELTING

4.4.1 FUNCTIONS OF THE CUPOLA

A cupola is intended to economically obtain molten metal ensuring castings with few defects, that is:

1. To produce hot and clean molten metal.
2. To produce molten metal with high fluidity.
3. To produce molten metal with proper chemical compositions.
4. To ensure economical and constant operation and ease of repair.

To fulfill these functions, it is necessary to:

1. Design the cupola with proper structure.
2. Select and use proper charge materials.
3. Establish and manage proper operating conditions.
4. Control the interprocess quality properly.

4.4.2 STRUCTURE OF THE CUPOLA

Figure 4.2 shows the basic structure of a cupola and the principal parts of a conventional cupola. The function of a cupola depends on the part below the charging door, which is divided from a functional point of view into the preheating zone, the melting zone, the superheating zone, and the well.

Metal charged through the charging door is first heated in the preheating zone by the combustion gas heat of coke and then is melted in the melting zone, followed by being subjected to superheating and tapped from the tapping hole through the trough. In a favorable operation, as shown in Figure 4.2, the furnace temperature is said to be 500–1000°C in the preheating zone, 1200–1500°C in the melting zone, and 1600–1800°C in the superheating zone; it is desirable that the tapping temperature be 1500–1550°C. The melting zone and the superheating zone are classified into the deoxidation zone and the oxidation zone from the viewpoint of combustion reaction. In cupola melting, the positions of these deoxidation and oxidation zones are important; they have a great influence on the properties of molten metal. When the oxidation zone is expanded to the top of furnace or when solid metal is put in a strong oxidizing atmosphere due to the lowering of the metal meltdown position, oxidation

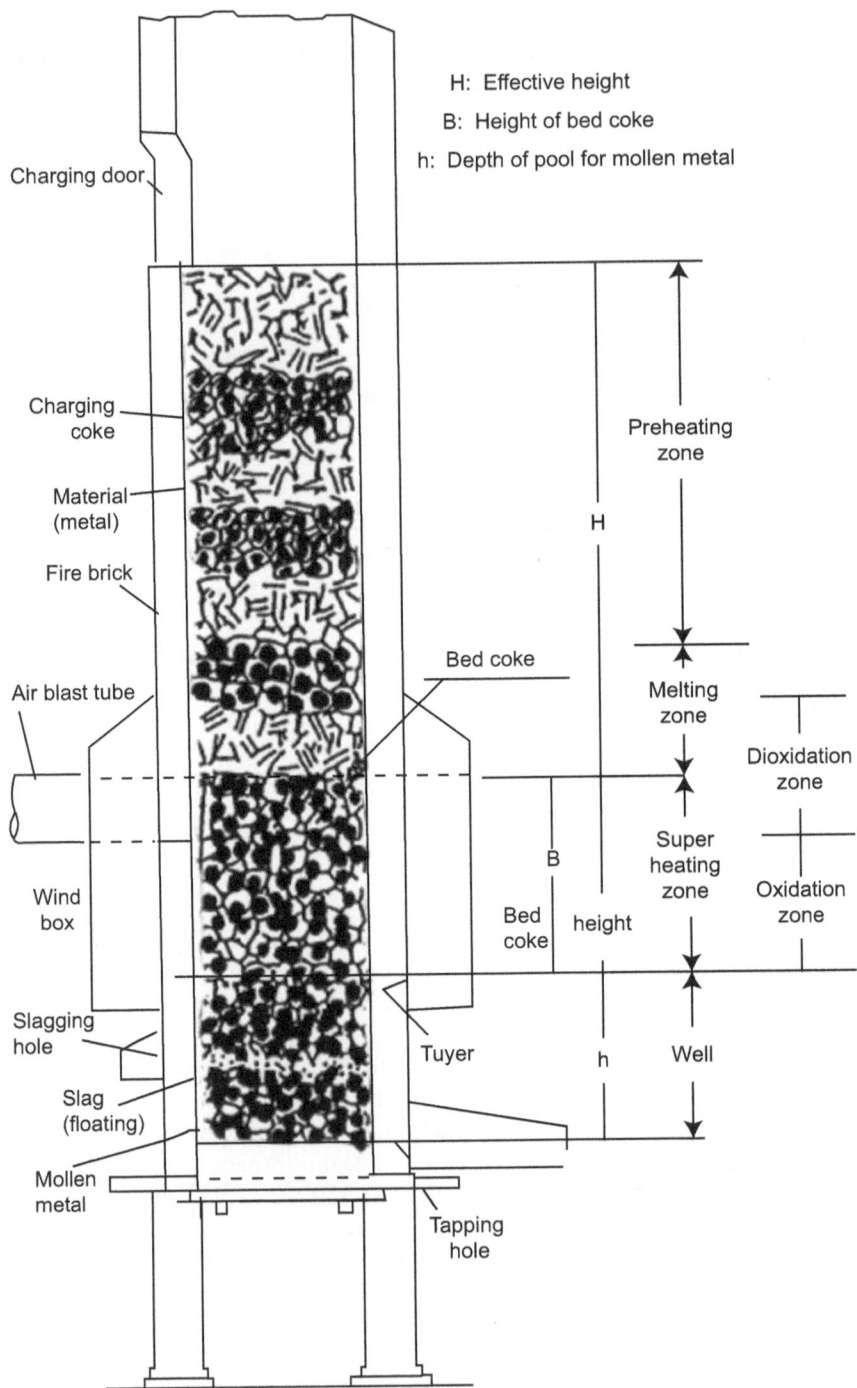

FIGURE 4.2 Structure of cupola and functional zones.

of the molten metal is accelerated, and the melting loss of Si is increased, which may cause the abnormal graphite form and defects such as shrinkage cavity, etc.

(1) Effective Height

The height from the tuyere (lower tuyere in the case of a multistage tuyere) to the lower end of the charging door is called the effective height, which is the most important part from a functional point of view. This part, which is a preheating zone where metal and coke are preheated by heat of combustion gas blown up from below and the moisture of the coke is evaporated, needs enough height, but excessive height may increase blast resistance and cause crushing of the coke at the time of charging. It is desirable that the height be 3.5–6.0 times as large as the inside diameter of the cupola.

(2) Tuyeres

Tuyeres are blasting ports for combustion air; this is an important part affecting the combustion of coke. Uneven pressure or quantity of air supplied from each tuyere leads to uneven combustion and heat generation of coke at the tuyere, causing oxidation melting in the low-temperature parts, thus generally lowering the temperature of molten metal. Much researches and many patents are related to equiblast, such as the form of the tuyere, wind box, buffer plate, etc. The ratio of the total cross section of tuyeres to the cross section of the furnace (the tuyere ratio) is 5 to 9 for a small furnace and 10 to 15 for a large furnace.

This is because, in the case of a large furnace, it is necessary to increase wind speed so that the wind reaches the inner part of the furnace. The number of tuyeres should be increased so that there are no dead points where combustion is insufficient; there should be 6 in the case of a small furnace and more with an increase in the size of the furnace. Tuyeres should be arranged horizontally on a plane at equal intervals surrounding the furnace.

(3) Wind Box

The wind box is intended to convert the kinetic pressure of air to static pressure to make equiblasting from each tuyere into the furnace. The wind box shall be designed so that the velocity head of air passing through the air blast tube is as small as possible to supply an equal quantity of air to each tuyere.

4.4.3 BEST OPERATING PRACTICES FOR CUPOLA

The cupola is the most common type of melting furnace used in the foundry industry. Heat is released by the combustion of coke. Carbon dioxide (CO_2) and carbon monoxide (CO) are released in the following reactions through the combustion of coke:

1. $C + O_2 = CO_2$ exothermic (heat generated)
2. $CO_2 + C = 2CO$ endothermic (heat absorbed)

In a divided blast cupola (DBC), reaction 2 (generation of CO) is suppressed by the introduction of a secondary (upper) row of tuyeres, about a meter above the primary (lower) tuyeres.

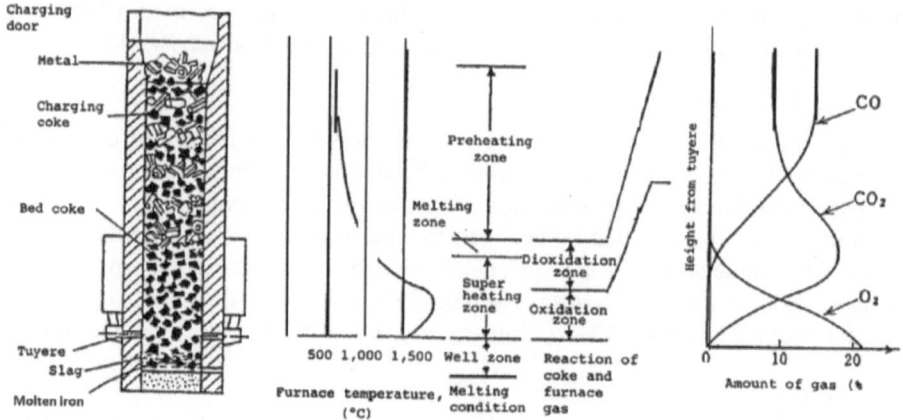

FIGURE 4.3   Combustion reaction and gas distribution in cupola.

4.4.4 Benefits of a Properly Designed and Operated DBC

- Coke consumption lowered 20–30%
- Increase in tapping temperature by about 50°C
- Higher melting rate for the same internal diameter (ID) of the cupola

Melting Rate of Cupola

- The melting rate of a cupola using low ash (14%) should be about 10 T/hr/ m^2, and for a cupola using high ash coke (30%), it should be about 7 T/hr/m^2.
- If the melting rate is lower, check the parameters of cupola blower.

Cupola Blower

The blast rate and pressure have an important influence on cupola performance.

4.4.5 Blast Rate

- Optimum blast rate is 375 ft^3/min/ft^2or 115 m^3/min/ft^2.
- Blower rating should be 15–20% more than the optimum blast rate in order to account for air losses in the pipeline.
- Do not overblow the cupola. A higher blast rate increases oxidation loss of iron and other elements like silicon and manganese.
- A lower blast air leads to lower metal temperature, slower melting, and higher coke consumption.

4.4.6 Blast Pressure

- Optimum blast pressure (P) = 0.005 (ID)2–0.0134 ID + 39.45, inch water gauge (WG). ID is in inches. (For conversion to kPa from inches-WG, divide by 4.0146.)

- Proper blast pressure is required to penetrate the coke bed. Incorrect air penetration adversely affects the temperature, carbon pickup, and the melting rate of the cupola. Incorrect and correct penetration of blast air is shown in Figure 4.4.

4.5 ENERGY EFFICIENCY BEST OPERATING PRACTICES GUIDE FOR FOUNDRIES

4.5.1 STACK HEIGHT

- Consider increasing the stack height of the cupola.
- A stack height between 16 and 22 feet (depending on its diameter) is necessary for heat exchange. Inadequate stack height leads to lower heat exchange and higher coke consumption.

4.5.2 WELL CAPACITY

- Do not increase well capacity more than what is desired.
- Every inch increase in the well depth reduces molten metal temperature by 4°C.

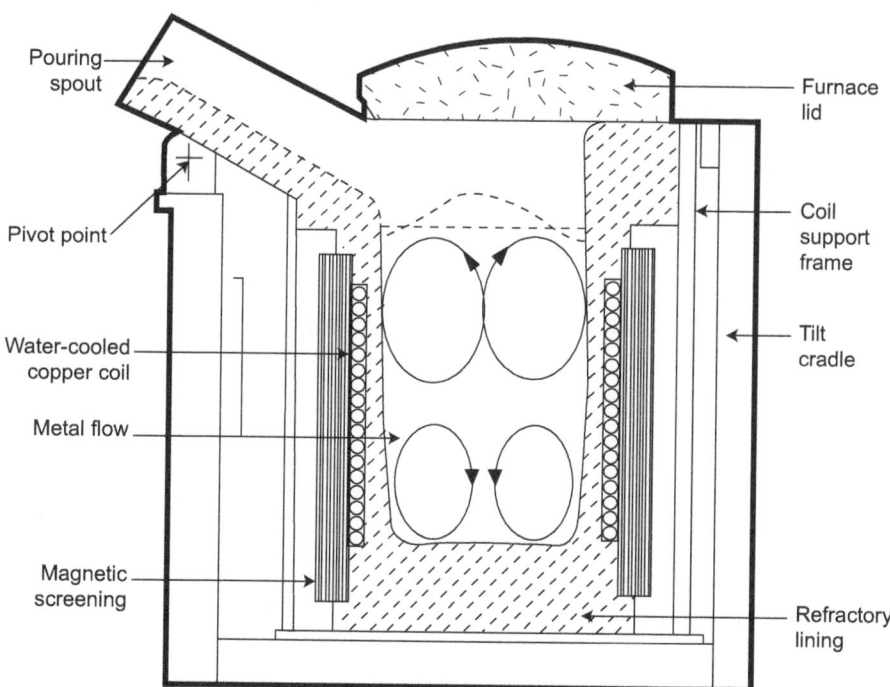

FIGURE 4.4 Schematic of induction furnace crucible.

- For an intermittently tapped cupola, the well capacity should match the capacity of the ladle.
- For a continuously tapped cupola, a minimum well depth, ideally of 300 mm, is usually sufficient.

4.5.3 REFRACTORY LINING

- Use IS 8 grade fire bricks for lining the cupola. For a double-brick-lined cupola, IS 6 grade bricks can be used in the rear side toward the cupola shell.
- Do not use cracked and corner-chipped bricks in the melting zone.
- Store the fire bricks in a shed to keep them dry.
- Fire clay should be soaked for at least 48 hours before use. This is required to develop its plasticity and the adhesive quality required for joining bricks.
- Provide proper tools, such as hammer, trowel (or karni), mallet to operators.
- Make a diameter gauge for the taphole.
- Provide a lighting arrangement inside the cupola during repair work.
- Bricks should set 15–20 mm away from the cupola shell. Pack this gap with dry foundry sand.
- Mortar applied on the bricks should not be layered thickly. It should be just enough so that the brick sets firmly on to the other. Excess spilled mortar is to be wiped off. Excess or thick layers of mortar are weak points through which molten metal can penetrate to the shell, causing hot spots on the shell.
- A 20/25 mm thick layer of ganister over the newly lined bricks is advisable.
- The cupola must be properly lined/repaired after each melt. Correct internal diameter (ID) needs to be maintained.
- Patching material can be used to repair the refractory lining if erosion/burn-back is less than 3 inches (75 mm). If erosion is greater, then use bricks for repair. To make 6 bags of patching material, mix the following:
  - 4 bags of refractory (or 3 bags of refractory + 1 bag of fireclay)
  - 1 bag of small grogs
  - 2 bags of bigger grogs (about ½ inch in size)
- Mix these materials in water (to check the consistency, the mix may be tested for stickiness on a vertical surface). Add 2 kg sodium silicate to the prepared mix, and mix it again. The patching material is now ready for use.
- A wooden bat (14 inch blade length with slight convex surface on one side) may be used for ramming the patching material during cupola repair.

4.5.4 BED PREPARATION

- Check the sand bed for correct sloping.
- Put dry firewood on the sand bed. Large pieces of wood should be avoided as this may prevent subsequent installments of coke to consolidate easily. Be careful that the sand bed is not damaged while placing the firewood.

- Select good-quality coke of proper size for the bed coke. Avoid too large and too small pieces.
- Weigh the bed coke and keep it in a separate heap. This is to be charged into the cupola in installments.
- Open all the tuyere covers. Put the first installment of coke inside and light up the firewood with cotton waste soaked in kerosene oil or LDO, as may be available. Plastic bags must never be used in place of cotton wastes or cotton rags. Some foundries use oil burners for ignition.
- Natural draught entering through the breast door and moving up to the cupola shaft helps ignition. Keep watching. As the coke gets ignited up to the (lower) tuyere level, add the second lot of coke. Go on adding in splits, as ignition progresses. Keep the last lot of coke in hand. This is to be charged after the ash blow-off and height bed measurement.
- Check through the tuyere to see that coke has ignited uniformly in an amber glow. With natural air ignition, the time should take 2–2½ hours. However, to expedite ignition a small (portable) blower may be used. Normally this is necessary in the rainy season when wood and coke contain more moisture.
- After the coke is satisfactorily ignited, close the upper tuyere covers. Keep the lower tuyere covers open. The taphole and slag hole should also be open. The fettling door is also to be kept open. Place guards in front of the tap hole and slag hole to arrest shooting sparks and coke pieces that can cause injury to people. Start the blower and blow off the ash for ½ minute strictly. This exercise cleans the furnace as well helps coke consolidation. After blow-off, open the tuyere covers. Close the breast door securely. Keeping the lower tuyeres (airflow) open, blow for another 2 minutes to ignite the coke fully. Shut the blower, open the tuyeres.
- Measure the bed height using a height gauge. The gauge is inserted from the top (charge door).
- The remaining last split of coke (kept for the bed) has to be put in before measuring the bed height.
- Before the start of metallic charges, the bed flux has to be changed on the bed.
- The flux (limestone in most cases) size should be ¾–2 inches for small cupolas.

4.5.5 Melting

Furnace operators should be in constant attendance. They must have proper implements and safety gear, such as:

- Goggles (blue).
- Hard gloves (leather).
- Poker for taphole).
- Crowbar.
- Small hammer.

- Bucker of water.
- Oxygen lancing gear (on standby).

4.5.6 Charge Material

- Charge material should be as clean as possible. Excess rust or dirt clinging to charge metal requires more flux material and coke and also leads to greater erosion of the cupola lining.
- Heavy (thick) sections of scrap should be avoided in the first five charges.
- No dimension of the charge material should be larger than ⅓ of the cupola ID.
- Correct weighing of charges is very important for proper and stable chemistry of the molten metal.

4.5.7 Cupola Operation

- Close the tuyere covers, plug the tap hole and slag hole. Allow some time for metallics to absorb heat. This is called soaking time. Generally, 10 minutes is sufficient.
- Switch on the blower and note the time.
- Look for droplets of molten metal through the tuyere peephole and note the time. Droplets should be visible approximately after 7–10 minutes of starting the blower.
- After blow on, the first tapping can be made within approximately 15–20 minutes.
- If the taphole gets jammed due to cold metal or any other reason, do not attempt to open it by hammer. Put some lighted cotton waste and charcoal at the taphole and apply oxygen with a lancing tube (MS pipe of 2 mm bore diameter) at regulated pressure.
- An oxygen cylinder (kept for lancing) must have a regulator fitted on it. Pneumatic pipe used for this purpose must match (bore-wise) with the lancing pipe. No leaks should be there.
- The oxygen melts the solidified metal at the taphole.
- In intermittent tapped cupolas, the slag hole should be opened after ⅔ tapping.
- Note the slag conditions, i.e., fluidity/viscosity or fluid. The ideal color should be bottle green.
- Constant vigil should be kept to ensure that at no time does the charge stack level fall. If this happens due to any reason, shut off the blower, open the tuyere covers, fill up the stack, close the tuyere covers, and the start blower.
- Maintain proper sequence of charging.
- Bridging or hanging inside the cupola should be taken care of immediately. A fall in stack level, bridging, or hanging affects the quality and chemistry of molten metal.
- Maintain a logbook recording every detail for every heat. Some of the important parameters (apart from the weight of each charge) that must be recorded in the sheet are bed height, weight of bed coke (this may vary with

the bulk density), bed light-up time, time taken for coke ignition, time taken and number of charges to fill up the stack at the beginning, blower on time, any interruption during melting, its cause, end time of melt, i.e., blower off, last tapping, drain-out, bottom door opened and bed dropped. All times must be meticulously recorded.

4.5.8 CLOSING OPERATION

- If the cupola has two levels of tuyeres, after the last charge, reduce air from the top tuyeres and increased it in the lower level. This can be done by means of the blast control valves.
- After the last tap, shut off the blower. Drain out the last metal, and open all the tuyeres.
- Remove bottom door props so that bottom drop-doors fall open.
- Before the pros (door support) are removed, be sure that no water is under the drop-door floor.
- The surrounding area should also be clear so that the operator removing the props can get away.
- After dropping, check for a clean drop through the tuyeres. Remove coke or slag by poking the tuyeres.
- Water cool drop-off to salvage coke.

4.6 BEST OPERATING PRACTICES FOR INDUCTION FURNACE

4.6.1 INTRODUCTION AND WORKING PRINCIPLE

The electric induction furnace is a type of melting furnace that uses electric currents to melt metal. The principle of induction melting is that a high-voltage electrical source from a primary coil induces a low voltage/high current in the metal or secondary coil. Induction heating is simply a method of transferring heat energy. Two laws that govern induction heating are: electromagnetic induction and the Joule effect.

High-frequency induction furnaces use the heat produced by eddy currents generated by a high-frequency alternating field. The inductor is usually made of copper in order to limit electric losses. The inductor is in almost all cases internally water cooled. The furnace consists of a crucible made of a suitable refractory material surrounded by a water-cooled copper coil. In this furnace type, the charge is melted by heat generated from an electric arc. The coil carries the high-frequency current of 500–2000 Hz. The alternating magnetic field produced by the high-frequency current induces powerful eddy currents in the charge, resulting in very fast heating. A typical schematic of induction furnace crucible is given in Figure 4.5.

There are two main types of induction furnace: coreless and channel. The coreless induction furnace has essentially replaced the crucible furnace, especially for melting high–melting-point alloys. The coreless induction furnace is commonly used to melt all grades of steels and irons, as well as many nonferrous alloys.

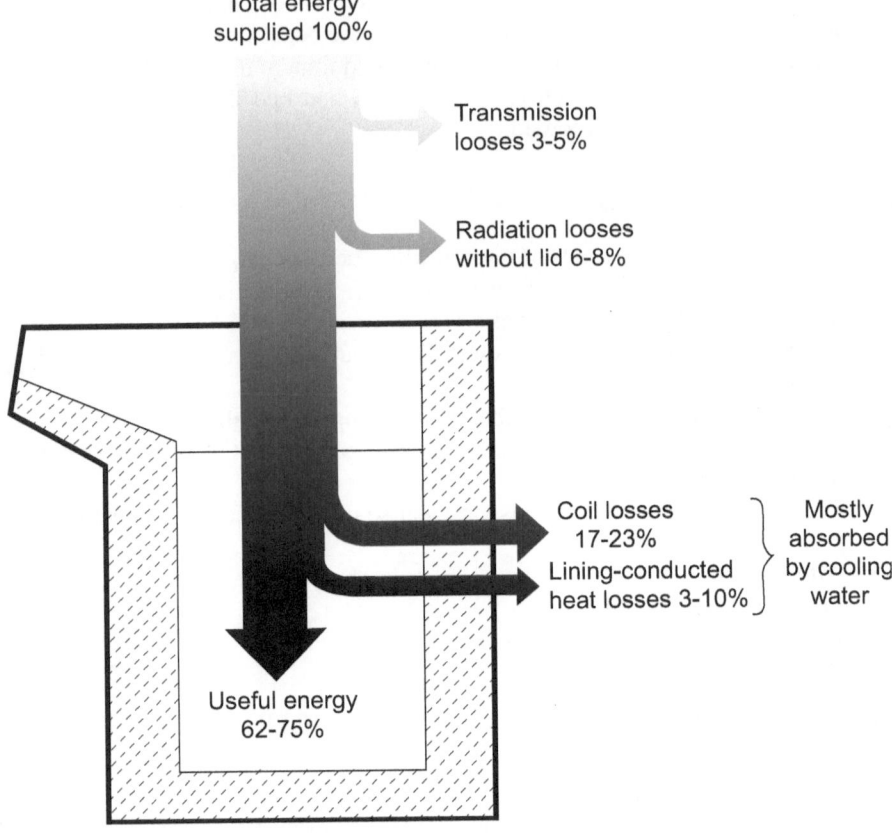

Total energy
supplied 100%

Transmission
looses 3-5%

Radiation looses
without lid 6-8%

Coil losses   Mostly
17-23%   absorbed
Lining-conducted   by cooling
heat losses 3-10%   water

Useful energy
62-75%

FIGURE 4.5   Sankey diagram of energy flow in induction furnace.

A modern coreless induction furnace can melt a tonne of iron and raise the temperature of the liquid metal to 1450°C using less than 600 kWh of electricity. Typically, specific energy consumption of a coreless induction furnace varies from 500 to 800 kWh/T depending on type and grade of casting. The overall efficiency of an induction furnace depends on many factors, such as the scrap charging system, furnace design, furnace cover, harmonics control, multiple-output power supply, and refractory.

4.6.2 Losses in Induction Furnace

The electrical energy required for heating 1 tonne of iron to 1500°C is 396 kWh. In the furnace, numerous losses take place that increase the specific energy consumption to above 500 kWh. The losses are thermal furnace losses, furnace coil losses, capacitor bank losses, convertor losses, and losses on the main side transformer. The losses are represented in Figure 4.6.

THE ENERGY REQUIRED TO MELT 1,000 KG OF IRON TO 1,550 DEG C IN A STEEL SHELL FURNACE

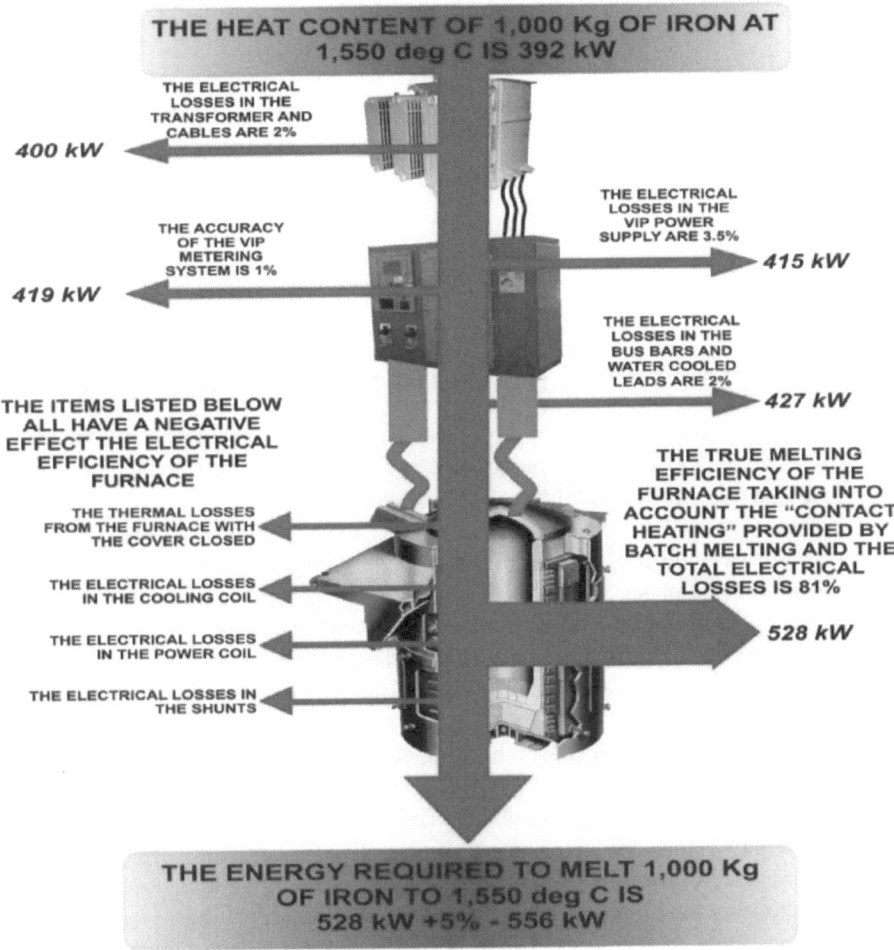

FIGURE 4.6 Energy enthalpy of iron.

In a typical induction furnace, the energy loss in equipment is between 100 and 130 kWh/T. The furnace efficiency is around 65–75%. With new developments in energy efficient coils, new refractory material, reduction of converter losses, and reduction in transformer losses, the state-of-the-art furnace equipment energy loss is reduced to 60–90 kWh/T. The new furnaces have an efficiency 81–87%.

Energy Enthalpy—Iron

The energy enthalpy of iron is described in Figure 4.7. A 1000 kg of iron is heated at 1550°C in a steel furnace.

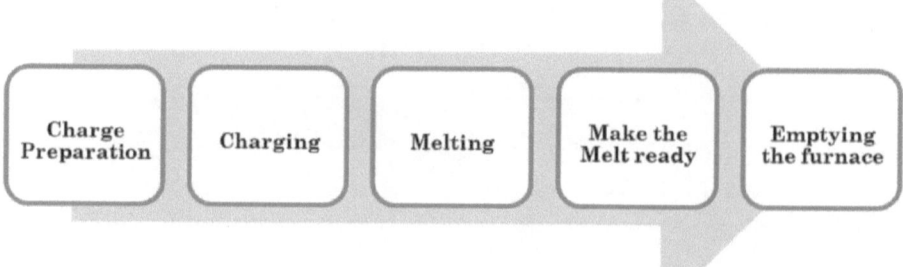

FIGURE 4.7   Stages of operation in induction furnace melting.

4.7 SIMPLE STEPS TO CUT DOWN THE ENERGY CONSUMPTION

1. **Operating procedures:** Replace the current equipment with equipment that is more productive and energy efficient. Investigate operational changes that also may produce significant cuts in your power use.
2. **Equipment utilization:** Maximizing your melting equipment efficiency— i.e., the way you use the equipment you have—can produce the greatest operational savings.
3. **Speedup of charging:** Ideally, charging should occur rapidly and should put the metal in the furnace as fast as or faster than the furnace is able to melt it under full power. This requires automated charging systems (except for smaller furnaces).
4. **Avoiding overcharging:** Overcharging scrap above the top of the furnace causes energy to be wasted.
5. **Using clean scrap:** Dirty charge materials also waste energy. For example, sand has two times the heat content of iron.
6. **Streamlined temperature measurement:** Frequently undershooting and overshooting the target temperature causes additional readings.
7. **Maintenance procedures:** Close attention to normal maintenance procedures can produce surprising energy savings—at minimum additional cost.
8. **Keeping refractory thickens as specified:** Thicker refractory means that the metal will be farther away from the coil, resulting in a higher current in the furnace coil and much greater current losses.

4.8 BEST OPERATING PRACTICES

Efficient operation of a coreless induction furnace depends primarily on implementation of good/best operating practices. The steps involved in the operation of an induction furnace are shown in Figure 4.8. Best operating practices under each stage are elaborated in the following section.

FIGURE 4.8 Vibrating feeder for induction furnace.

4.9 CHARGE PREPARATION AND CHARGING

- The raw material must be weighed and arranged on the melt floor near the furnace before starting the melting.
- Charge must be free from sand, dirt, and oil/grease. Rusty scrap not only takes more time to melt but also contains less metal per charging. For every 1% slag formed at 1500°C, energy loss is 10 kWh/T.
- The foundry returns, i.e., runner and risers, must be tumblasted or shot blasted to remove the sand adhering to it. Typically, runner and risers consist of 2–5% sand by weight.
- Keeping the exact weight of alloys ready. As alloys are very expensive, their proper handling reduces not only wastage but also time lost in alloying.
- The maximum size of single piece of metal/scrap should not be more than ⅓ of the diameter of the furnace crucible. The right size avoids the problem of bridging. Moreover, each charge should be about 10% of crucible volume.
- There should be no or few sharp edges, particularly in the case of heavy and bulky scrap, as this may damage the refractory.
- The furnace should never be charged beyond the coil level, i.e., charging the furnace to its capacity. It should be noted that as the furnace lining wears out, the charging may slightly increase.
- A proper charge sequence must be followed: bigger-sized metal first, followed by smaller sizes. Gaps must be filled by turnings and boring.
- Limit the use of baled steel scrap and loose borings (machining chips).
- Use charge driers and preheaters to remove moisture and preheat the charge. Vibrofeeders for furnaces are equipped with a vibrating medium, and they could be fuel fired to preheat the charge and remove oil/grease. This is shown in Figure 4.9.
- Avoid the introduction of wet or damp metal in the melt; this may cause an explosion.

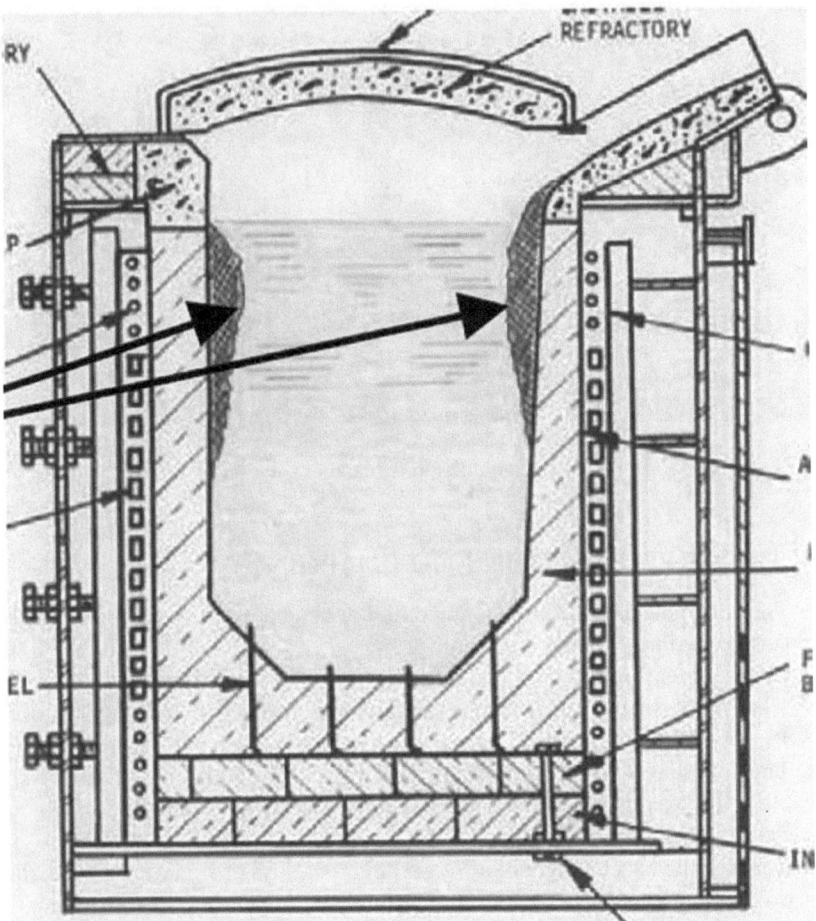

FIGURE 4.9   Slag buildup near furnace crucible neck.

4.10 MELTING AND MAKING THE MELT READY

- Always run the furnace at full power. This not only reduces batch duration but also improves energy efficiency. For example, in a 500 kg, 550 kW furnace when run at full power, the melt may be ready in 35 minutes, but if the furnace is not at full power, it may take over 45 minutes.
- Use the lid mechanism for the furnace crucible; radiation heat loss accounts for 4–6% of input energy. For instance, a 500 kg crucible melting at 1450°C with no lid cover leads to radiation heat loss of up to 25 kWh/T.
- Avoid buildup of slag on furnace walls, as shown in Figure 4.10. Typical slag buildup occurs near the neck, above the coil level where agitation effect is less. The quantity of flux used for slag removal is important. Typically, flux consumption should be less than 1 kg/T of metal.

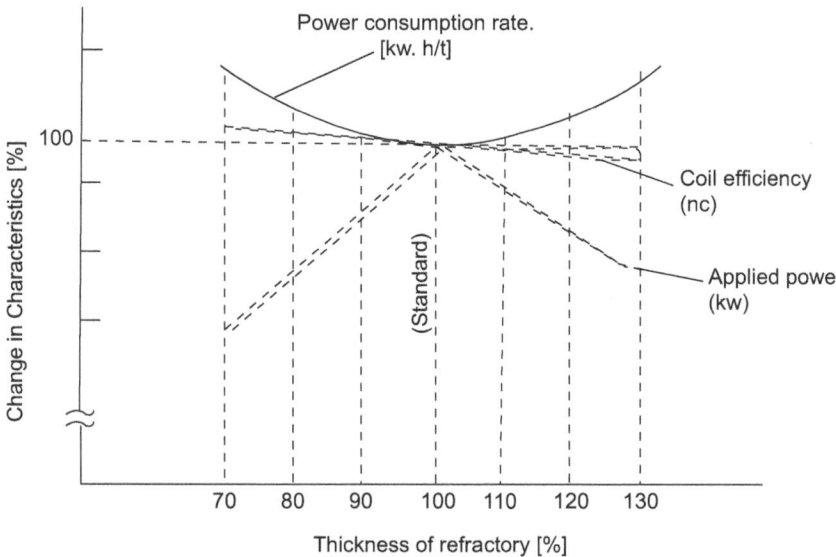

FIGURE 4.10 Lining thickness vs. power consumption rate.

- Proper tools must be used for deslagging. Use tools with flat heads instead of rods or bars for deslagging; the right tool is more effective and takes much less time.
- Process control through the melt processor leads to fewer interruptions and typically reduces interruptions by 2–4 minutes. The spectrotesting lab must be located near the melt shop to avoid waiting time for chemical analysis.
- Avoid unnecessary superheating of metal. Superheating by 50°C can increase furnace specific energy consumption by 25 kWh/T.

4.11 EMPTYING THE FURNACE

- Plant layout plays an important role in determining the distance travelled by molten metal in the ladle and the temperature drop.
- Optimize the ladle size to minimize heat losses, and empty the furnace in the shortest time.
- Plan melting according to molding capacity. Metal should never wait for mold; rather, the mold should be ready before the metal.
- Use a ladle preheater. Using molten metal to preheat the ladle is very energy intensive and expensive.
- The quantity of liquid metal returned to the furnace must be as little as possible.
- Use a glass-wool or ceramic-wool cover for the pouring ladle to minimize temperature drop.
- Minimize plant breakdowns by implementing a planned maintenance schedule.

4.12 FURNACE LINING

- Select the correct lining material.
- Do not increase lining thickness at the bottom or sidewalls. Increases in lining mean reducing the capacity of the furnace and increasing power consumption. The effect of increasing or decreasing lining thickness on power consumption rate is shown in Figure 4.11.
- Do not allow the furnace to cool very slowly. Forced air cooling helps in developing cracks of lower depth, this helps in a faster cold start cycle. Cold start cycle time should be ideally not more than 120% of normal cycle time.
- Coil cement should be smooth, in straight line, and of a thickness of 3–5 mm.
- While performing, the lining ensures that each layer is not more than 50 mm. Compaction is better with a smaller layer.
- Consider use of preformed linings.
- Monitor lining performance.

4.13 ENERGY MONITORING AND DATA ANALYSIS

- A separate energy meter for the furnace must be installed.
- Monitor energy consumption on heat-by-heat basis. Analyze them in correlation with production data to arrive at the specific energy consumption of the furnace on a daily basis.
- Any peak or valley in data must be studied and investigated in conjunction with tapping temperature and quantity of metal tapped.
- Energy monitoring is the first step for achieving energy saving.

FIGURE 4.11   Raw material storage.

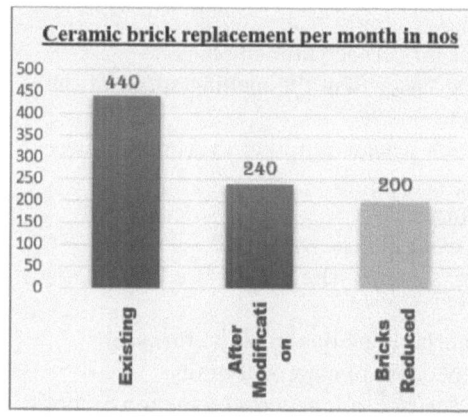
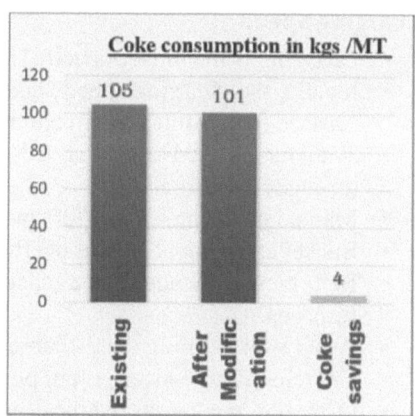

FIGURE 4.12   Refractory lining comparison and fuel consumption trend.

4.13.1 OTHERS

- Effective raw material storage is important for optimum performance of the furnace. For example, bundled scrap stored on a mud floor leads to dust and moisture pickup. See Figure 4.12.
- The temperature and flow rate of coil-cooling and panel-cooling waters must be monitored.
- The panel must be checked on a weekly basis, and cleaning must be done on a monthly basis.
- Check the condition of fins in the cooling tower, and clean the fins on a monthly basis.

4.14 ENERGY COST REDUCTION

There are two areas for energy cost reduction:

1. Operating procedures
2. Maintenance procedures

4.14.1 OPERATING PROCEDURES

Equipment Utilization

Higher equipment utilization means shorter holding times.

- Longer holding times mean greater energy consumption.
- Shorter holding times optimize Kwh/T of metal poured.
- Dual- or Tri-Trak equipment boosts equipment utilization.
- Typically, utilization of Mono-Track is about 60%, Dual-Trak is 85%, and Tri-Trak is above 95%.

Charging Speed

- Charging is the most impactful factor for furnace utilization.
- Ideally, the charging speed should be faster than the melting speed of the furnace (under full power condition).
- The charge needs to be added in such a way as to get maximum power throughout the melt cycle.
- Manual charging is okay for small furnaces.
- Bucket charging is okay if overhead clearance is available.
- Belt conveyors/automated charging systems should be used for large furnaces.
- Avoid overcharging (cold charge material lying on top of the furnace).
- Overcharging reduces output power because of electrical limits.
- Overcharging causes overheating of furnace components on the top part of furnaces, leading to energy losses.
- Overcharging also affects the functionality of the furnace lid.
- Never allow the bath to go to the molten stage when cold charge material still has to be added. This is to avoid wet or damp material causing eruptions from the furnace when moisture comes in contact with liquid metal.
- In other words, add cold charge on top of nonmolten material in order to drive out moisture with heat from the top bath.

Temperature Measurement and Sampling

- Maintain temperature control without undershooting or overshooting the target temperature. More time consumed for adjustment results in underutilization.
- Shifting from the melting power to the holding power at the appropriate stage helps in temperature control.
- Time consumed for sampling needs to be optimized with proper setup and process.

Optimize Pouring

- Too much tapping time calls for metal reheating, causing wastage of electrical energy.
- Proper setup (like large ladles) and practices to tap metal reduce tapping time.
- Pour at the coolest possible temperature. Overshooting the pouring temperature by 10% boosts heat losses by 30%, wasting energy.
- The appropriate pouring temperature also reduces refractory wear and additive alloy losses.

Schedule or Control Usage of Melting Power

- Optimize metal need and keep the furnace at full power to save on demand charges.
- If possible, melt for fewer days per week and more hours per day.
- Doing so reduces holding time/power and also increases refractory life due to reduced hot/cold cycles.
- Fewer hot/cold cycles mean higher refractory life and hence more throughput per lining.
- In turn, this means fewer sintering cycles, thereby saving energy.

4.14.2 MAINTENANCE PROCEDURES

Furnace Lid

- About 75% of heat loss is in the form of radiation loss and 25 % of heat loss via conduction through refractory and floor.
- Effective utilization of the lid reduces radiation losses.
- Radiation losses increase exponentially with metal temperature.
- A 10% rise in pouring temperature results a 33% rise in radiation losses.
- Look for possible lowering of the peak pouring temperature.

4.15 ENERGY SAVING PROPOSALS AND OPPORTUNITIES

4.15.1 CASE STUDY IMPLEMENTED IN COIMBATORE FOUNDRY CLUSTER BY MODIFICATION OF EXISTING CUPOLA FURNACE WITH LAMINAR AIRFLOW CONTROL

Measure

The existing blowing pipelines in the cupola furnace can be modified into a funnel shape. With this modification, air circulation in the combustion chamber is uniform. If the air circulation is even, it results in the oxygen enrichment needed to meet the coke blasting requirement. Hence the melt rate is increased and uniform. This can reduce overall coke consumption by 4–5%. Another important advantage is uniform air circulation in the cupola combustion chamber, reducing wear and tear on the ceramic bricks.

Summary

The blast rate and pressure have an important influence on cupola performance. Proper blast pressure is required to penetrate the coke bed and proper combustion. Incorrect air penetration affects the temperature and melting rate of the cupola.

Principle

Oxygen can also be used as a cost saving tool by allowing decreases in charged coke, partially substituting lower-cost anthracite for more expensive foundry coke, replacing pig iron with less expensive charge metals, and decreasing ferroalloy requirements. Oxygen enrichment is defined as a method of increasing the oxygen content of the blast air from 21% to a higher level. This is normally achieved by adding pure oxygen to the air while reducing the nitrogen content of the air-oxygen mixture.

The theoretical air required to burn the coal is 7.1 kg of air for 1 kg of coal.

The proposed design system reduces brick wear and tear so that the same bricks can be used for next two melt processes. Ramming mass cements can be used for rework in place of the brick replacement process.

- Blower delivery pressure: 850–1000 mm WC
- Air volume supplied by blower: 1500 cubic m/hr
- Delivery pipe length: 10–12 feet

Background

MS Sri Ramkrishna Industries is a medium-scale cupola-based foundry unit located near Goldwin's, Coimbatore. The average monthly production is around 45–50 MT/month. This unit has installed one 30 inch cupola furnace for producing the casting components. A detailed furnace study has been conducted with an external agency, MS Rajpreeth Industries, in this unit during the fourth quarter of 2015. After detailed technical discussions of how to improve the energy efficiency of the existing furnace, the following points were arrived at:

1. The blower delivery line should be connected in the center of the combustion chamber with the shape of funnel.
2. A V-type air separator should be fixed to distribute uniform air in the combustion chamber.
3. Tuyere pipelines need to be modified as curves instead of as straight lines

The existing average melt rate per hour is 1560 kg/hr, and coke consumption per tonne is around 105 kg.

Benefits of Implementing the Measure

- Air circulation to the combustion chamber is uniform.
- The melt rate is increased to 9.6%.
- Coke consumption is reduced by 4%.
- Ceramic brick replacement during lining work is reduced by 55%.
- Yield is improved of quality product.

The present melt rate per hour is 1685 kg/hr, and coke consumption per ton is around 101 kg.

Cost–Benefit Analysis

Sl. Number	Description	Units
1	Type of measure	Medium
2	Coke savings per annum	2200 kg
3	Ceramic bricks reduced/year	2000
4	Cost savings from coke—Rs. 36/kg	79,200
5	Cost savings from bricks—Rs. 45/brick	90,000
6	Annual cost savings	1,69,200
7	Investment cost	1.35 lakhs
8	Payback period	10 months

FIGURE 4.13 Cupola inside after melt.

FIGURE 4.14 Cupola inside after lining.

Conclusion

After modifying the combustion chamber inlet pipeline as per the proposal, the melt rate was increased to 10% due to a proper blast rate. Coke consumption is also reduced to 4%, and refractory lining work reduced to 50%. The average cost savings per annum is around Rs. 1.69 lakhs.

CO₂ reduction per year is 6.82 MT.

4.15.2 ENERGY SAVINGS THROUGH MATERIAL CHARGING

Performance optimization of induction furnace (use of small pieces of MS scrap for charging)

FIGURE 4.15   Energy saving through material charging.

4.15.3 ENERGY SAVINGS THROUGH MATERIAL CHARGING

Replacement of old induction melting furnace with new EE induction furnace

FIGURE 4.16   Energy saving through material charging.

4.15.4 Energy Savings through Lid Mechanism

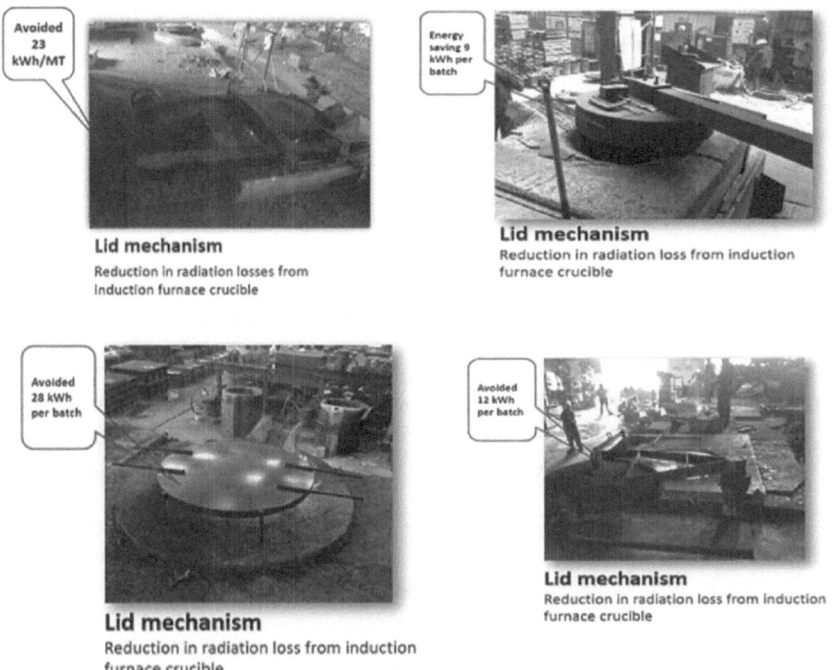

FIGURE 4.17   Energy saving through lid mechanism.

4.15.5 Energy Savings in Furnace Cooling Tower Measure

The cooling water system in the induction furnace is a very important element in removing the heat generated by the furnace coil. Around 180–200 kWh/MT are used for removing the heat produced by the coil in induction-based foundry units. In general, if the cooling water system is not well managed, the coil may be damaged, and more water may be consumed.

Summary

In the existing system, this unit was installed with a 240 TR cooling tower in the cooling system of Inductotherm Corp.'s Tri-Track furnace with a capacity of 1750 kW.

The cooling tower has a 10 kW pump for raw water circulation, and a 5 HP motor is used for cooling the fan motor. Under normal operating conditions, both the fan motor and the pump are running continuously irrespective of the temperature of the raw water.

At present in the cooling tower were installed a raw water sump, one RTD (resistance temperature detector) PT100 (temperature sensor), and one temperature controller. Hence, the temperature controller shuts down the fan motor when the water temperature reaches 36°C and turns it on again when the temperature reaches 40°C.

Due to these modifications, the fan motor goes idle for about 6 hours a day. The average energy savings per day is around 20 kWh.

Benefits of Implementing the Measure

- Reduced specific energy consumption in the cooling water system
- Reduced water consumption
- Fan motor running only between 36 and 40°C

Conclusion

By installing an RTD with a temperature controller in the cooling tower, the specific energy consumption is reduced to *20 kWh/day*. The payback period is less than 2 months. *CO₂ reduction per year is 5.0 MT.*

4.15.6 Energy Savings through Best Operating Practices

Measure

Specific energy consumption of the liquid metals depends on the raw material charge pattern, the quality of the raw material, and the power setting of the induction furnace. During the energy study process in this unit, the raw material charging pattern

Cost–benefit analysis

1	Average energy consumption per day by the existing fan motor	76 kWh
2	Present energy consumption	56 kWh
3	Energy savings per day	20 kWh
4	Energy savings per year @ 300 days	6000 kWh
5	Total energy cost savings per year(Rs. 8.0/kWh)	Rs. 0.48 lakhs
6	Equipment capital investment cost	Rs. 0.08 lakhs
7	Return on investment period	2 months

Energy cost = Rs. 8.0/kWh.

FIGURE 4.18   Temperature sensor with controller.

and quality of the raw material are found to be good. The power setting of the furnace is found 600 kW, and the capacity of the furnace is 750 kW. Due to aging of the furnace (20 years old), the power setting is reduced to minimize thyristor failure. During the energy audit period, the power setting increased to 630 kW from 600 kW.

Summary

In the present melting process (600 kW) setting, the specific energy consumption of the liquid metals is around 626 kWh/MT. After increasing the setting to (630 kW), the specific energy consumption of the liquid metals is around 617 kWh/MT. Nearly 1.4% energy savings (9 kWh/MT) can be achieved through adopting such best practices.

Benefits of Implementing the Measure

- Melt time is reduce by up to 5 minutes.
- Metal output is increased by up to 750 kg/day.
- The SEC for liquid metals is reduced by up to 9 kWh/MT.

Cost–benefit analysis

1	Energy savings per MT	9 kWh/MT
2	Melt time reduction per heat	5 minutes
3	Additional heats per day	0.5 heats
4	Energy savings per day	126 kWh
5	Energy savings per year @ 300 days working	37,800 kWh
6	Extra liquid metals per day	750 kg
7	Increase in the melting capacity per month	18,750 kg
8	Total energy cost savings in per annum*	Rs. 2.835 lakhs
9	Yield improvement	3.57%

*Energy cost = Rs. 7.5/kWh.

Existing Energy Consumption @ 600 kW Setting

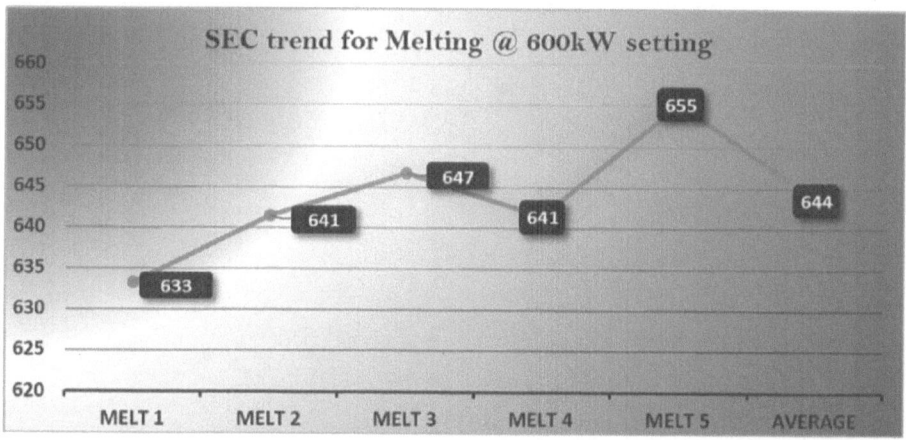

FIGURE 4.19 SEC trend for melting at 600 kW setting.

The graph in Figure 4.19 depicts the SEC for liquid metals in a specific duration. Based on the melting history, the average SEC for liquid metals is around 626 kWh/MT.

During the melt study process in the furnaces, it was found that the hot metal is maintained around 100 kg in the crucible. As a result, the melting performance such as melt time is maintained around 80 minutes and for pouring around 20 minutes.

Parameters	FG iron	SG iron
Melt time in minutes	80 minutes	80 minutes
Tapping time in minutes	20 minutes	20 minutes
Tapping temperature	1500–1510°C	1500–1520°C
Pouring temperature	1480–1410°C	1480–1420°C

Observations

- CI borings are received from the supplier mostly in the form of chips.
- Steel scraps are received as bundles.
- Hot metal is maintained at around 100 kg.
- Coil-cooling raw water inlet temperature is around 35°C, and the outlet is around 52°C.
- Coil-cooling pump water flow is around 290 L/min with velocity of 1.44 m/s.

An energy study was carried out in the furnace for various melts. A three-phase power quality analyzer was fixed in the furnace panel, and further study was conducted.

Present Energy Consumption @ 630 kW Setting

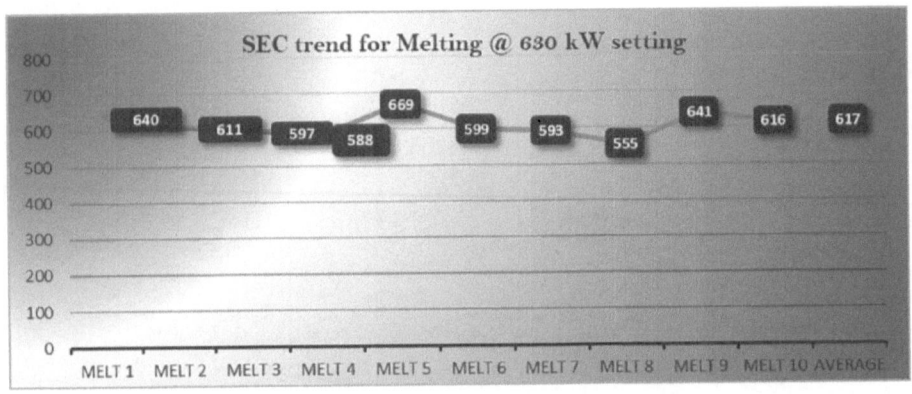

FIGURE 4.20   SEC trend for melting @ 630 kW setting.

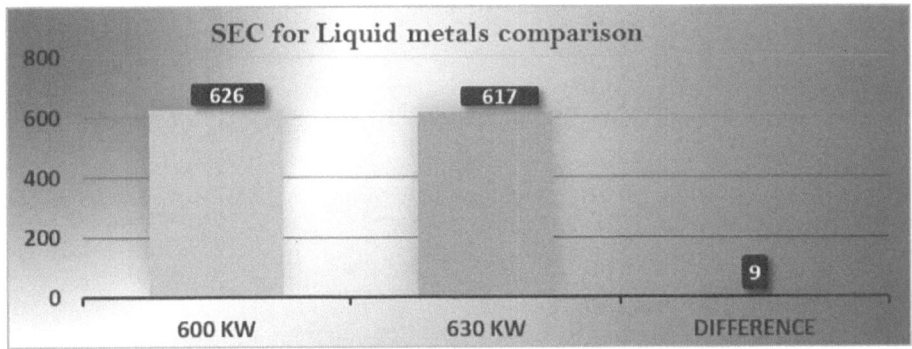

FIGURE 4.21   SEC for liquid metal comparison.

Conclusion

Average energy consumption for the liquid metals @ 630 kW is around 617 kWh/MT. Energy savings is around 9 kWh/MT.

Expected CO_2 reduction per year is 31.5 MT.

REFERENCES

[1] Best operating practices guide book for foundries—developed by TERI under GEF-UNIDO-BEE project. http://sameeeksha.org/brouchres/BOP-Guide-Oct2015.pdf
[2] Best Practices in Melting, Inductorthem India private limited.
[3] Afgan, N. H. and Beer, J. M. (1974) Heat Transfer in Flames, Scripta Book Co., Wiley, New York.
[4] Beer, J. M. (1974) Methods of calculating radiative heat transfer from flames in combustors and furnaces. In Heat Transfer in Flames, Scripta Book Co., Wiley, New York.
[5] Hewitt, G. F. Shires, O. L., and Bott T. R. (1994) Process Heat Transfer, CRC Press.
[6] Hottel, H. C. (1974) First estimates of industrial performance; the oneâ€"gasâ€"zone model re-examined. In Heat Transfer in Flames, Scripta Book Co., Wiley, New York.
[7] Hottel, H. C. and Sarofim, A. F. (1967) Radiative Heat Transfer, McGraw-Hill, New York.
[8] Kern, D. O. (1986) Process Heat Transfer, McGraw-Hill, New York.
[9] Truelove, J. S. (1983) Furnaces and combustion chambers. In Heat Exchanger Design Handbook, Hemisphere Publishing, New York.

5 Energy Conservation Opportunities in Electrical Systems

L. Ashok Kumar and Gokul Ganesan

CONTENTS

DOI: 10.1201/9781003203810-5

LEARNING OUTCOMES

At the end of this chapter, the reader will be able to understand:

- Energy conservation opportunities in electrical distribution systems.
- Energy saving proposals in electrical motors.
- Energy saving proposals in lighting systems.
- A real-time case study with saving calculations.

5.1 ENERGY SAVING PROPOSALS IN ELECTRICAL SYSTEMS STUDY

5.1.1 INSTALL HARMONIC FILERS AND SUPPRESS WAVEFORM DISTORTION

Present Status

DC drive with thyristor controls is employed in the production areas. Controlled rectifications with variable firing of the thyristor chop off the sinusoidal waveform and highly distort it, resulting in multiple fundamental waves of voltage and current, called voltage and current harmonics.

A thyristor load of 20% or more of the total load utilized or more than 5% of the present level develops harmonics of considerable magnitude, necessitating installation of adequate harmonic filters.

Presently the thyristor load constitutes 41.41% of total connected load, and average power demand constitutes 44.68% of the total plant average demand. This produces approximately 20% total harmonic, resulting in harmonic losses of 2–3% of total energy drawn.

It is extremely difficult to theoretically assess harmonic losses. But it has been established that a total harmonic distortion of around 20% in an induction furnace results in a loss of 3–3.5% of the system energy.

Modification Proposed

Conduct a detailed harmonic study at the three transformer secondary panels. Based on the results, install harmonic filters on the secondary buses.

Energy Conservation Potential and Benefits

The expected harmonic losses on a conservative estimate will be around 2% of annual energy drawn, i.e., 2,5 lakhs kWh, which can be saved by installing harmonic filters. The cost of the saving would be Rs. 8.74 lakhs/annum.

Investment and Payback Period

The estimated filter installation cost, including study cost, is Rs. 11.5 lakhs for a 600 KVAR filter, which will be paid back in 16 months.

5.1.2 Install High-Tension Voltages Stabilizer for the Incoming Power Supply to Obtain a Constant Voltage

Present Status

At present, four transformers rated 750 LVA, 22,000/440 V each are installed to serve the low-tension (LT) power of the plant's electrical system. These transformers are divided into two groups, with two transformers running in parallel in each group. The voltages' fluctuation from the electricity board is very bad; it varies from 360 to 390 V during peak load periods and from 420 to 430 V during off-peak load hours and holidays.

The existing transformers have off-load changers.

The major power-consuming loads are furnaces with nichrome heating elements with a connected load of 1400 kW. The nichrome heating elements are very sensitive to voltage variation. The heating power delivered by these elements is roughly proportional to the square of the heating current.

Apart from this, the heat conversion efficiency of those elements is adversely affected by lowered voltage.

Test readings taken on one furnace show the power consumed at an average voltage of 405 was 41kW; with a lower voltage of 385 V, power consumed was 37.5 kW.

Based on this data, the energy consumed by the furnace for a particular load at normal voltages works out to 123 kWh, wasting 31 kWh/cycle, i.e., 20%. This is a very conservative estimate taking only an hour of extended time for heating.

The furnaces are rated for 415/440 V, at which they work at optimum efficiency with minimum consumption of energy.

Modification Proposed

Provide a voltage stabilizing facility to overcome the voltage fluctuation problem.
Two options are available for voltages stabilization:

1. Provide an onload tap-changing facility to the existing transformers.
2. Provide a voltage stabilizer on the 22 kV incoming side.

Energy Conservation Potential and Benefits

1. Stabilized voltage at the proper level increases the performance of the furnace, as the nichrome heater works at optimum efficiency.
2. Yield from them improve production rate increases considerably, making best use of the installed capacity of furnaces.
3. Rejects are reduced, resulting in fewer breakdowns.
4. Electrical equipment damage, especially that related to voltage variation, is reduced.
5. The annual savings in energy assuming present production level works out to 5,16,540 kWh costing Rs. 18,51,500.

Investment and Payback

The cost of a 2 MVA, 22,000 V stabilizer and installation works out to Rs. 26,30,000, which will be paid back in 17 months.

5.1.3 REALLOCATE THE LOADING PATTERN OF DIFFERENT TRANSFORMERS

Present Status

Presently there are eight transformers in the plant, and most of them are underloaded.

Modification Proposed

Transformers operate with maximum efficiency when they are loaded at 50–70%. The transformers are loaded far below the rated KVA, i.e., 359 against 1250 KVA. It is recommended to install a lower capacity transformer of 750 KVA. The transformer (750 KVA) can be used for this purpose.

The annual savings in reallocating the loading pattern of the transformer in the plant is Rs. 7410.

5.1.4 Operate the Two Transformers in Parallel to Reduce Transformer Losses

Present Status

Power is received from the electricity board, and three 1000 KVA, 33 kV/433 V transformers are installed for stepping down to 433 V for plant distribution. Each transformer feeds its own PCC (power control center), and the facility is available to run the transformer in parallel.

Now the transformer is run independently, and the loads in them are not balanced. The load on the TR 2 and 3, which were in service, was monitored for 24 hours.

These transformers have their maximum efficiency at 25–50% of loading. As per monitoring, transformer 3 was loaded around 50%, and transformer 2 is loaded at less than 25% of their respective rated capacities, both operating outside their maximum efficiency ranges.

Modification Proposed

Run two transformers in parallel to operate them at their best efficiency ranges and save copper loss.

Energy Conservation Potential and Benefits

By operating the transformers in parallel, 4035 kWh of energy are saved per year, for a cost saving of Rs. 14,204.

Investment and Payback Period

No investment is required, and hence the payback is instant.

5.1.5 Lay Additional Runoff Cable from Substation to Boiler Plant Feeder End Panel

Present Status

The cable feeder of the boiler plant from the substation is overloaded. The size of the cable feeder is not adequate to carry the current. This not only threatens the life of the cable but also incurs considerable I^2R losses. Based on the present loading, the I^2R losses incurred work out to 8404 kWh, costing Rs. 28,020.

The current carrying capacity of a 120 mm^2 cable is 156 A (considering a derating factor of 0.72) with the present method of installation, whereas the present load of the cable is 162 A. This results in an annual loss of 5603 kWh.

Modification Proposed

It is proposed to lay additional runoff cable from the substation to the boiler plant feeder end panel. This will reduce the current to half of the present value. The I^2R losses will also be reduced.

Energy Conservation Potential and Benefits

Since the current in the identification cable is reduced, the annual cable losses are also reduced, thus saving energy 2801 kWh, for a cost saving of Rs. 9861.

Investment and Payback

The total estimated cost of laying cable is Rs. 6502, which will be paid back in 8 months.

5.1.6 INSTALL ENERGY SAVER FOR WELDING TRANSFORMER

Present Status

At present, six 30 kVA welding transformer sets are available. Each machine normally works for 4 hours/day.

Modification Proposed

It is suggested to install an energy saver for the welding sets. Welding sets are not always on-load, although they are permanently kept switched on. During the idle time, the welding sets consume power at a lower level with a very low power factor.

The energy saver is ideal for both types, i.e., transformer and rectifier, to save energy. The electronic unit automatically switches off the set by cutting off supply to the conductor when the welding set is not in use for more than 5 seconds of the factory preset time interval. The time interval can be increased or decreased if the requirement is indicated at the time of ordering. To continue welding, the operator needs to touch the electrode to the job so that the electronic unit switches on the welding set by energizing the coil of the contactor. A system bypass facility is also provided.

Energy Conservation Potential and Benefits

1. The savings obtained by providing energy savers are Rs. 76,860.
2. There are a reduction in maximum demand and a consequent reduction in maximum demand charges.
3. The power factor of the system is improved.

Investment and Payback Period

The investment is Rs. 29.600, which will be paid back in 5 months.

5.1.7 LOAD MANAGEMENT OF ELECTRICAL ENERGY REQUIREMENTS BETWEEN IN-HOUSE AND IMPORTED POWER

Present Status

The plant has two electrical energy sources, one from the electricity board and the other from the in-house generation facility comprising two 590 KVA and one new 1000 KVA diesel generator sets.

The average cost of energy works out to Rs. 3.52/kWh.

The overall cost of energy produced in-house is Rs. 2.97/kWh, and the combined cost of total energy utilized works out to Rs. 3.45/kWh.

The high cost of electricity board energy is due to the high contracted demand, 75% of which is charged as minimum payable. Even the recorded MD (maximum demand) is high because the average load factor is only 0.57. This situation will be greatly improved when the new generator is run as a base load set.

Modification Proposed

Immediately after commissioning the new generator, in order to ascertain its reliability while running continuously, it is suggested to reduce the contracted demand to run this generator as a base load set to produce 46,80,000 kWh a month (at around 650 kW). The other two 590 KVA generators are to be run to 300 kW for 150 days to produce 10,80,000 kWh each a year.

With the new generator commissioned, which should produce at least an average of 468,000 kWh average per month, the maximum demand will never exceed 1500 KVA, even with an average low-load factor of 0.50.

The existing maximum demand controller is to be set to give warning when the maximum demand tends to approach 1400 KVA, so that the electrical staff can cut off some nonpriority loads or increase the load on the DG (distributed generation) sets, as usually these demand periods last only momentarily.

Energy Conservation Potential and Benefits

No reduction in the energy consumed is suggested; only the balancing of the total energy requirement between two sources is recommended to achieve cost reduction for the electrical energy requirements.

The annual energy saving of energy works out to Rs. 44,48,586.

Investment and Payback Period

The total investment is Rs. 19 lakhs, and it will be paid back within 5 months.

5.1.8 Improve the Load Factor to Achieve Optimum Cost of Electrical Energy

Present Status

The plant receives power from the electricity board at 33 kV.

The average energy received from board is 6,43,097 units per month, the maximum being 8,91,630 units and the minimum 2,60,640 units.

Maximum demand established was 1842 KVA with a minimum of 1116 KVA, the average being 1509 KVA. The established maximum demand reached only 74.45% of the contracted demand, and hence the contracted demand can be reduced to take advantage of the improved load factor.

The maximum load factor was 0.75, and the minimum 0.32 and average was 0.58. The maximum consumption and demand occurred when the load factor was low, i.e., 0.67. Scope exists to improve the load factor to 0.75. Considerable savings in the monthly recorded demand, resulting in savings in the cost of electrical energy imported, can be realized.

Modification Proposed

Reduce the maximum demand to 1950 KVA.

Set a maximum demand in the beginning of each month based on the estimated energy consumption and a load factor of 0.75. With the use of maximum demand controller, the preceding method can be implemented.

Energy Savings Potential and Benefits

The cost of improved power is escalating day by day. Hence every effort has to be made to reduce the demand by increasing the load factor. The annual cost of savings in maximum demand that can be achieved by this method works out to Rs. 4,31,404.

Investment and Payback Period

The investment required for additional relays wiring, etc. is only around Rs. 20,000.

5.1.9 Optimization of Energy Cost by Using Fuel Additive in HSD

Present Status

The plant has two electrical energy sources, one from the electricity board and the other from the in-house generation facility comprising two 500 KVA, 415 V diesel generator sets.

A study of the electricity energy consumption for the preceding 12 months reveals the following:

1. The average energy received from SEB is 1,51,448 kWh per month, the maximum being 95,590 kWh during May. The calculation shows the projected cost of energy with the latest tariff for the pattern of consumption during the preceding 12 months. The average cost of energy works out to Rs. 4.94/kWh.
2. The average cost of energy produced by the DG set is Rs. 6.28/unit as given in the calculation.
3. The maximum recorded demand is 892 KVA during the month of March, and average demand is 830 KVA.

Modification Proposed

An additive is recommended, namely Micromix 1000 D to HSD, which will reduce the fuel consumed by the DG sets. The savings achieved will be 5–6%. We have only taken a conservation figure of 3% in our calculation.

Energy Conservation Potential and Benefits

The annual savings by adding fuel additive are Rs. 15,198.

Investment and Payback Period

The investment for fuel oil additive is Rs. 8510, which will be paid back in 7 months.

5.1.10 Install 115 KVAR Power Factor Correction Capacitors on Feeder End Panels

Present Status

Power factor correction banks of 200 KVAR and 75 KVAR are installed at plants 1 and 2, respectively.

The reading taken by us at the feeder end indicates that the power factor varies from 0.216 to 0.91.

Modification Proposed

Install a 115 KVAR capacitor at the feeder end distribution board, with capacitance value as shown in the calculation, with suitable control switches.

Energy Conservation Potential and Benefits

By adding the capacitors to the feeder ends, the feeder power factor improves, reducing the current in the feeder cable and also lessening the strain on the cables. This arrangement is also useful for DG supply loads.

Since the feeder switches handle currents with a good power factor (0.94), they operate without strain. The switching off is cut out very quickly, and the contact life is prolonged.

Since the current in the cables is reduced, the annual cable losses are reduced, saving 12,029 kWh in energy costing Rs. 59,425.

Investment and Payback Period

The total investment is Rs. 16,100, which will be paid back in 3 months.

5.1.11 Install Low-Tension Voltage Stabilizer for the Distribution Power Supply to Obtain a Constant Voltage

Present Status

At present, one transformer of rating 500 KVA is installed in the plant to step down voltage from 11 kV to 433 V. The maximum demand recorded is 343 KVA. The voltage fluctuation is too much. It varies from 360 to 480 V during peak load periods. The existing transformer has off-load tap changers. The taps can be adjusted only with the power shut off, which affects the production.

Modification Proposed

It is recommended to install an automatic voltage LT stabilizer of 500 KVA ratings on the LT side with necessary protection in order to maintain a steady LT voltage.

Energy Conservation Potential and Benefits

1. Reduced rejects as a result of fewer breakdowns
2. Reduced breakdown of electrical equipment, especially the motor, which is sensitive to voltage variations
3. The annual savings in energy after installing the voltage stabilizer work out to 82,973 kWh costing Rs. 2,65,513 as shown in the backup calculation.

Investment and Payback Period

The cost of a 500 KVA LT stabilizer and installation works out to Rs. 5,20,700, which will be paid back in 24 months.

5.1.12 Use Additive in Lube Oil for Compressor and DG Sets

Present Status

Presently additive is not added to the lube oil for the compressor and DG set. The annual consumption of lube oil is 5313 L.

Modification Proposed

It is recommended to add a Teflon-based additive to the lube oil (Nulon–E25).

Energy Conservation Potential and Benefits

1. Life of the lube oil is increased by 250 hours.
2. Annual energy savings by adding lube oil additive is Rs. 20,700.

Investment and Payback Period

The investment for the additive is Rs. 41,500, and the payback period is 24 months.

5.1.13 Reduce DG Set Frequency to 48.5

Present Status

In the factory, the following operating frequencies are:

1. DG sets: 49.7–49.9 Hz.
2. TNEB grid: 47.8–48.5 Hz.

Any reduction in frequency will result in a reduction in the power consumption.

Modification Proposed

It is suggested to operate the DG set between 48.5 and 49.0 Hz.

Energy Conservation Potential and Benefits

We expect a minimum of 4% in savings in power for 2% reduction in frequency.

5.1.14 Reduce Frequency for AHU

Present Status

The AHU blower has a variable frequency drive. The motor was running on 50 Hz continuously without any temperature-based control for the VFD.

Modification Proposed

Reduce the frequency to 45 Hz, which will cause the motorized valve at the chilled water circuit to fully open, and this will reduce the load on the chiller.

5.1.15 Replace Defective Capacitors

Present Status

The capacitors connected at the substation (100 KVAr) are not giving the rated KVAr output and are also consuming power.

The maximum allowable power consumption of capacitors is 3–5 w/KVAr.

The capacitors at the substation draw 100 A, which is more than the norm.

There is good potential to save energy by replacing the identified faulty capacitor with new low-loss dielectric capacitor.

Modification Proposed

It is suggested to replace the identified capacitor with energy efficient low-loss capacitor banks.

Energy Conservation Benefits

The annual savings is Rs. 0.52 lakhs. The investment required is Rs. 0.35 lakhs, which will get paid back in 8 months.

Investment and Payback Period

The estimated investment is Rs. 0.35 lakhs, and the payback period is 8 months.

5.1.16 Optimization of UPS System

Present Status

Two UPS of rating 1000 KVA and 400 KVA are connected in the plant. Maximum load is less than 1000 KVA. The company is paying annual maintenance contract of 3 lakhs for 400 KVA UPS system. There is a 3.7 kW fan for 400 KVA UPS.

Modification Proposed

It is recommended to use only 1000 KVA UPS, and the 400 KVA can be kept as standby. The efficiency of the 1000 KVA is higher than that of the 400 KVA.

Energy Saving Potential and Benefits

1. There will be savings of Rs. 3 lakhs since there are no payments for the annual maintenance contract.
2. The annual saving by optimizing the UPS system is Rs. 4,17,440.

Investment and Payback Period

There will not be any investment for the proposed system, and hence payback in instant.

5.1.17 Provide Captive Generation Facility to Obviate Losses due to Electricity Board Power Interruption

Present Status

The company receives power at 33,000 volts from Maharashtra State Electricity Board. The voltage is stepped down to 415 V by three transformers with "on-load tap changers" for the plant power distribution.

The contracted demand with the electricity board is 2700 KVA. This is quite high as the highest maximum demand in the 10 months of processing was only 1340 KVA_{in} May 1996. This is not indicative of normal demand as the load factor during this month was low.

The electricity board has revised its tariff in July 1996, steeply hiking the energy cost from Rs. 2.42/kWh to Rs. 3.07/kWh and that of demand cost from Rs.123/KVA to Rs. 170/KVA. With this hike, given the pattern of consumption by the company, the unit cost works out to Rs. 3.53.

Since all the incoming transformers are provided with on-load tap changers, the LT voltage is maintained fairly constant at 415 V.

Though this eliminates process disturbances due to voltage fluctuations, of late there had been too many power interruptions with duration ranging from 15 minutes to 6 hours. This has seriously affected their production. Considerable loss in production has resulted, and if evaluated in terms of cost, the value assumes phenomenal dimensions.

As furnished by the company, the optimum total loss of production for six months from December 1995 to May 1996 was a 154.31 MT contribution to Rs. 92.58 lakhs. This works out to an annual loss of almost Rs. 2 crores.

The imported power cost is rising day by day. The present cost of Rs.3.53/kWh is without any fuel cost adjustment; this may eventually be levied, pushing up the unit cost further. It will not be too bad a guess to expect the unit cost to reach Rs. 4 by the year-end.

Modification Proposed

With the electricity board energy cost steeply rising and reliability coming down to the point of seriously affecting production, captive generation is becoming more and more indispensable for almost all plants in India. This is more pronounced in the case of continuous process industries like Cosmo Films Ltd., which cannot tolerate even momentary power interruptions, let alone unscheduled, prolonged power shutdowns. The gap between electrical power demand and generation in India is widening day by day, leaving no hope for any improvement in the situation. Even noncontinuous engineering and service industry plants have resorted to installing captive generation plants to augment their electrical energy requirement. It is but a logical step for a company to immediately install a captive generation plant to augment at least 50% of its requirement.

Benefits and Cost Reduction Operation

The cost/unit of in-house generation works out to Rs. 3.15, which is less by Rs. 0.378/kWh, and at 5 lakhs units/month, the energy cost saved would be Rs. 22.68 lakhs/year. The annual saving by reduction in demand works out to Rs.12,27,825, as shown in the annex. The total savings in the energy bill is Rs. 34,96,000, with the elimination of a production loss of Rs. 1,85,16,000, the total amount of savings would be Rs. 2,20,12,000.

Investment and Payback Period

The total cost of investment, for the installation of gensets and all auxiliaries works out to Rs. 2,75,00,000, which will be paid back in 15 months.

5.1.18 Replace Motor-Generator Sets with Thyristor-based Welding Sets

Present Status

Presently, motor-generator (MG) welding sets are used in the plant. The M-G sets convert the AC to DC supply.

The modern trend is to use a thyristor rectifier for converting AC to DC supply. The efficiency of the MG sets is low (80–82%) compared to thyristor rectifiers (95%). MG sets are rotating equipment, and the efficiency of rotating equipment is low compared to nonrotating equipment.

The no-load power consumption of the MG sets vary from 1. 80 to 2.2 kW, which is high compared to a thyristor rectifier.

Modification Proposed

Replace the MG sets with thyristor-based welding sets.

Investment and Payback Period

The estimated investment is 1 lakh, and the payback period is 33 months.

5.1.19 Reduce Voltage by 5% at the Transformer

Present Status

Power is received from the State Electricity Board (SEB) at 11 kV and is stepped down to 415 V. Many motors are running at less than their rated capacity.

Modification Proposed

It is proposed to reduce the distribution supply voltage by 5% by adjusting the tap changer in order to obtain a voltage of 395 V.

Energy Conservation and Benefits

The reactive power drawn by inductive loads is proportional to the square of the applied voltage. When the motors are not fully loaded, they do not require the full air gap flux to transmit the power to the rotor and hence the voltage. Hence the applied voltage can be marginally reduced. By reducing the voltage, the reactive power and the magnetic losses in an induction machine like a motor can be reduced from 5 to 10%, resulting in lower consumption of total power by about 4%. By reducing the voltage, the annual energy saved may work out approximately to about 2,00,000 kWh, costing Rs. 8,32,000 at Rs. 4.6/kWh.

Further this will improve the power factor of the total system, resulting in the reduced maximum demand and distribution and other losses.

5.1.20 Use Super Kerosene for the DG Sets

Present Status

For the DG sets in the plants, diesel is used as the fuel.

Modification Proposed

It is proposed to use super kerosene as the fuel for the diesel generator sets.

Benefits and Savings

The super kerosene can be used in the diesel generator sets without any modification in the fuel injection system. It has a calorific value of 11,000 kcal/L, compared to 10,800 kcal/L of diesel. Also, the flash point of super kerosene is 34°C, compared to 36°C of diesel, and hence no special precautions are necessary in its use. The super kerosene costs Rs.15/L, compared to the price of diesel, which is Rs. 20/L. The total amount spent on the fuel is reduced, which gives good savings in the cost of production of the energy. There is a net savings of Rs. 5/L.

In a plant with an annual consumption of 100 kL, the savings comes to $100 * 1000 * 5 =$ Rs. 5,00,000.

5.1.21 Install Energy Saver for DG Sets

Present Status

There are two DG sets of 500 KVA rating.

Modification Proposed

It is proposed to install magnetic resonators to save fuel consumption. In the DG sets, the fuel savings potential is 2–3%.

Investment and Payback Period

The estimated investment is RS. 96,000, and payback period is 19 months.

5.1.22 Improving the Performance of the DG Set by Fine-tuning

Present Status

The present specific energy generating ratio (SEGR) of the 500 KVA DG set is 2.2 at 60% loading.

This is low.

This indicates that the engine needs fine-tuning in the following areas:

1. Air/fuel adjustment
2. Injection value timing adjustment
3. Servicing

Modification Proposed

It is suggested to carry out:

1. Proper value timing adjustment.
2. Cleaning of filter.
3. Air/fuel ratio adjustment.

Investment and Payback Period

The investment required for improvement is Rs. 60,000, and the payback period is 6 months.

5.1.23 Install Electrical Meters at Different Sections of the Plant (Where Meters Are Not Presently Available) for Monitoring and Targeting Energy Utilization in the Plant

Present Status

Presently no accurate metering system is available in the plant so as to record the electrical energy utilization at various parts of the plant, such as crushing, milling, production, lighting, and welding works. At certain locations (e.g., the VFD control room and old PH room), the meter currently in service needs recalibration.

Modification Proposed

It is suggested that a separate maximum demand ammeter and digital energy meter be provided at all the power distribution control centers, like mills 1, 2, 3 and sections A, B, C, so as to assess the power consumption and line losses at various process locations. In addition, a LT feeder arrangement with the required energy meter and ammeters may also be provided for lighting, welding, and air handling units. The meter that requires recalibration should be taken out and put back online after recalibration as early as possible.

Energy Conservation Potential and Benefits

These steps will help to monitor the electricity consumption in the plant closely and thus help to achieve an energy savings of 0.5–1% of the total energy supplied.

Investment and Payback Period

The net total investment is Rs. 55,000, and the payback period is 2 months.

5.1.24 Provision of Capacitors across the Terminal of the AC Welding Transformer

Present Status

The repair and reconditioning works in the plant require welding work. Six welding sets are in service. Out of these, four units are AC of the 20 KVA AC welding type, and the rest are of the 15 kW DC type.

The operation of the AC arc welding sets not only leads to a poor power factor of the system but also causes energy losses due to their poor efficiency (around 50–60%) three-phase sets. In the case of two-phase AC welding sets, the efficiency is still low (30%).

Modification Proposed

To avert the adverse effects caused by the poor power factor operation of these AC arc welding sets, it is suggested that a shunt capacitor (6 KVAR) may be provided directly at the terminal of the four AC welding transformer in the plant.

Investment and Payback Period

The estimated investment value is Rs. 14,400, and payback period is 7 months.

5.1.25 REENGINEERING OF THE POWER DELIVERY NETWORK

Present Status

The load pattern in the plant varies from minute to minute and is influenced by such factors as variations in supply frequency, voltages, imbalances in current, the presence of harmonic and fluctuation loads. The aged equipment, control system, and UG cable system have dust accumulated over them, and this lead to increased losses in the plant. Being a continuous process plant, all the essential machines work around the clock.

Being a continuous process industry, the load factor is high (above 90%). This load factor depends on the availability of the raw material, viz. sugarcane for crashing. Since the working of this plant depends on the availability of sugarcane, the load factor is not a constant. The point to be noted in this regard is that a low load factor leads to higher losses in the plant.

Modification Proposed

The reengineering the entire power delivery network in the plant may be considerable as one of the best solutions in the reduction of the present losses in this system to the level around 3–3.5%. The net saving in terms of energy and money per annum will be around 0.5% (3 lakh units and Rs. 12 lakhs, respectively).

Before undertaking the reengineering of the present power delivery network, it is preferable to measure the actual losses in the plant by installing a network of accurate energy meters at all the designated locations. With the help of a single software program, the losses can be calculated with the aid of a computer. Once the actual measurement is carried out, the real condition of energy consumption in the plant can be known. With this knowledge, an appropriate reengineering of the power delivery system with good design practices in the plant can be well planned and executed with a payback period of 2–3 years.

In connection with this, it is suggested that, out of the existing four 2 MVA distribution transformers and 27 runs of LT cables, two transformers and 15 runs of the LT cables may be considered for reengineering, and they may be shifted nearer to their load centers. In the instant case, they may be shifted to the old powerhouse control room. While carrying out this remodification process, separate LT feeders may be provided for lighting, welding and AC plants.

It is worth mentioning here that, at another sugar factory, 2 MVA transformers are located very close to the load enters, and separate LT feeders provide for lighting loads. Thus it strengthens the need for the present proposal.

While carrying out the reengineering of the power delivery system, an automatic meter reading system is now proposed for this plant. Such a system will read automatically the energy recorded at the energy meters at various parts of the plant and transmit the data to a central database for effective use and better understanding of the energy distribution in the plant.

5.1.26 Replacement of Single Welder Type (Two-phase Connection) 400 V, 20 KVA Oil-Cooled Welding Transformer with Three-phase Oil-Cooled Welding Transformer or with Modern Rectifier Types

Present Status

Two 20 KVA 400 V single-welder-type welding transformers are employed in the plant. The efficiency of the single-welder-type welding plant is 30% or less.

Modification Proposed

The efficiency of double-welder-type three-phase 400 V 20 KVA welding transformer is 50% and above. From an energy saving point of view, the double-welder-type welding transformer is preferable. Hence it is suggested that the existing single-welder-type welding transformer may be replaced with a doubled-welder-type welding transformer or be operated for a minimum number of hours per day until replacement is effected.

Investment and Payback Period

The estimated investment is Rs. 30,000, and the payback period is 30 months.

5.2 ENERGY CONSERVATION IN ELECTRICAL MOTORS

5.2.1 Install Automatic Star-Delta Controller for Underloaded Motor with Six Terminal Configurations

Present Status

At present, the automatic star-delta controller is available for most of the motors having six terminal windings; i.e., each winding is independently connected to the terminal. The starter initially connects the motor winding in star mode, and when the motor reaches 80% of its rated speed, it automatically connects the windings in the delta. During starting, each winding gets reduced to single-phase voltage to limit the starting current. Many motors are underloaded, drawing power at a low power factor. At this condition, the efficiency of the motor is very low, with high magnetic losses. When the motor is loaded at less than 35% of its rated output, the power factor can be improved by reducing the voltage.

Modification Proposed

1. It is suggested to install an auto-delta controller for motors, with an automatic load sensing facility. This facility senses the load continuously and switches automatically to and from star-delta depending on the load on the motor, saving magnetic losses. This is an add-on unit for the existing star-delta starter.
2. It is observed that many of the motors are partially loaded. There is a saving of power of 58% if the motor is operated on the star position. The existing starter need not be changed; only by disconnecting the timer of the starter is the motor made to run in star mode. At a later date when the load increases, the starter can be set to operate in the normal delta position.

For all the motors, when loading is less than 35%, this practice can be adopted; the average savings per motor will be around 25% of present load on the motor.

Investment and Payback Period

The cost of the timer for automatic star-delta controller and installation works out to Rs. 12,000, which will be paid back in just 1 month.

5.2.2 INSTALLATION OF SOFT STARTERS FOR SELECTED EQUIPMENT

Present Status

At present, there are three instrument air compressors and four process air compressors: one is continuously running, one compressor is running with variable load, and one is a standby.

Energy Conservation Potential and Benefits

1. The soft starter applies appropriate voltage, and, depending on the load on the motor, reactive current is kept at an optimum-level reduction, lessening magnetic and other constant losses.
2. The current-limiting option provides controller torque and soft start capability.
3. Soft starter is ideal for energy saving in motors that are operating at partial or cyclic loads.
4. Voltage control improves.
5. Sudden starting jerks, with shocks to the bearings, couplings, gears, and driven equipment, are avoided, increasing the useful life cycle.
6. Current is continuously maintained at the optimum level, which reflects a reduction in KVA demand.
7. Maintenance of the motor and driven equipment is also reduced.

The saving obtained by the installation of soft starter works out to Rs. 1.6 lakhs.

Investment and Payback Period

The investment required is Rs. 3.9 lakhs, which will be paid back in 29 months.

5.2.3 REPLACE PRESENT FLUID COUPLING DRIVES BY VARIABLE FREQUENCY DRIVES

Present Status

Some drives are provided with fluid couplings to have a smooth start of the load and in some cases to have speed variation. Power consumed by these couplings is considerable, as the heavy couplings run at full speed.

Modification Proposed

Replace the present fluid coupling drives by variable frequency drives.

Energy Savings Potential and Benefits

The fluid coupling drives are inefficient, and power loss in the couplings increases as the speed increases. The variable frequency drives control the speed by frequency variation. These drives are very efficient at around 98%; the speed can be varied smoothly, and the life of the drive and driven equipment increases.

By providing VFD for the six drives, an annual energy savings of 7020 kWh, costing Rs. 24,710, can be achieved.

Investment and Payback Period

The cost of the VFD with configuration works out to Rs. 73,600, which will be paid back in 36 months.

5.2.4 Install Variable Speed Drives for Cooling Water

Present Status

The blower is used for furnaces cooling in the melting zone. Blower air discharge is controlled according to the requirement by damper. The cooling air requirement increases gradually, and the blower is operated at full load capacity after 2–3 years of campaign life.

At present, there are two blowers of 50 HP each in plants 1 and 2 and one blower of 125 HP in plant 3.

Modification Proposed

It is suggested to install variable speed drives (VSDs) in these blowers, so that the output of blower discharge can be controlled by changing the motor speed.

Savings

The annual saving by installing VSDs in the cooling blowers is Rs. 27,10,826, and investment is Rs. 32,80,000. The payback period is 15 months.

5.2.5 Provide an Interlock System for the Mill Section

Present Status

At present there is no interlock system for the mill and the auxiliary drives; when the mill is idle, some of the auxiliary loads that are manually controlled are running.

Modification Proposed

We suggested providing an interlocking system for the main mill driven so that the auxiliary loads are cut down when the main mill is stopped. The saving in power by providing the interlock system is around Rs. 8 lakhs/year.

5.2.6 Replace V-belt Drive by Flat Belt Drive

Present Status

At present the motor drives are required to be equipped with V-belts.

Generally, V-belts are used for the transmission of power from the drive to the driven due to the fear of slippage of a flat belt. The transmission efficiency of a V-belt drive is in the range of 87–88%. The wedging action of a V-belt involves an irreversible energy loss as each belt is continuously wedged into the pully groove and pulled out again. The power losses in a V-belt are typically dissipated in the form of heat, which in turn reduces the belt life.

Modification Proposed

It is recommended to replace the V-belt drives by flat belt drives for one air compressor.

The performance of drives driven by the flat belt depends on the quality of belt and pulley. Today flat belts are available that are manufactured with state-of-the-art technology for application in all areas of industry.

Energy Conservation Potential and Benefits

1. The transmission efficiency of flat belt drive is 98%.
2. Flat belt drives are very efficient and save 4–10% of energy consumed by the equipment.
3. The service life of the flat belt drive is four times longer than that of V-belt due to reduced wear on the belt and pulley.
4. Once the flat belt has been correctly installed, it is maintenance-free and guarantees constant power transmission during the entire service life.
5. These types of drive ensure smooth running as the tension is uniformly distributed over the entire belt width.

Investment and Payback Period

The investment for replacing a V-belt to flat belt drive is around Rs. 10,000, which will be paid back in 18 months.

5.2.7 Replace Metal Cooling Fans of Indicated Motors by Plastic Fans

Present Status

Metal cooling fans have been replaced with plastic fans in many motors. Still, some of the motors in the plant are fitted with heavy cast iron cooling fans. The metallic fans are heavy and over the course of a year develop pitting due to erosion and allow material buildup and consume more power.

Modification Proposed

Replace the metal cooling fans of the motor by plastic cooling fans.

Energy Conservation Potential and Benefits

Plastic cooling fans are lighter and also prevent material buildup, and hence they consume less power. As the mechanical power consumed by the rotor assembly is constant irrespectively of the load on the motor, the energy saving resulting from this modification will be around 2% of the rated kilowatts of the motor. The total annual saving obtained by changing to plastic fans is 343 kWh, i.e., Rs. 1695.

Investment and Payback Period

The estimated cost of replacing metal fans by plastic fans is Rs. 850, which will be paid back in 6 months.

5.2.8 Replace Underloaded Motor with Correct-Sized Energy Efficient Motor

Present Status

Some of the motors in the plant are running at very low load; either the original sizing is very high, or the estimated load is high. A standard induction motor runs at optimum efficiency only when loaded to 75% of its rated capacity. At lesser loads, efficiency drops very badly, as mechanical, stray, and magnetic losses are constant and not dependent on the loading. In the case of old motors, efficiency is low. Even at full load, efficiency is very low, and at lower loads, the motor efficiency falls off steeply.

Modification Proposed

Replace the underloaded motor with a correct-sized energy efficient motor.

Energy Conservation Potential and Benefits

1. The motor starting current and torque are modified, and hence starting jerks are avoided. This increases the life of gears, couplings, and driven equipment.
2. The flat top in the efficiency curve ensures savings at part loads also. Efficiency reaches the maximum at about 60% of the load. Hence the motor runs at optimum efficiency.
3. Higher power factor yields reduce line current and save distributed losses.

By replacing the identified underloaded motor with a correct-sized energy efficient motor, the energy saving achieved will be 1795 kWh, costing Rs. 8865.

Investment and Payback Period

The total investment for the energy efficient motor works out to be Rs. 9500, and it will be paid back in 13 months.

5.2.9 Install Variable Speed Drives for Motor of Pump Running with Control Valves in Highly Throttled Condition

Present Status

In the chemical plant, about ten pumps are running with the flow regulated by control valves. These valves are in a highly throttled condition.

Controlled flow of liquid by throttling results in considerable waste of energy across the valves. At 20% opening, three to four times the actual power required for the particular flow rate is wasted. Also, to accommodate this wastage of power, a higher horsepower motor is installed.

Modification Proposed

It is suggested to install variable frequency drives to vary the speed of the motor to regular flow. The power drawn by the pump is proportional to the cube of the flow rate.

Hence if the pump speed is regulated, the power required will be almost proportional to the cube root of the flow. If the flow rate is cut by 5%, the power required will be only 12.5% of rated power. The control valve has to be kept fully open, and the speed of the motor has to be regulated automatically by the VFD.

Since the power required is drastically reduced, the motor rating can also be reduced.

Since the investment is high, it is recommended to install one VFD for the pump, as a case study, monitor the performance, and then take up other installation.

Energy Conservation Potential and Benefits

By installing these variable speed drives, a saving in energy of 24,480 kWh can be achieved, costing Rs. 76,867.

Other benefits are as follows:

1. For smaller motors, lower losses
2. Smooth starts without high starting current and less strain on the electrical system, gear boxes, pumps, etc. along with lower maintenance
3. Prolonged motor life
4. Automatic regulation of the speed to suit the load requirement, using the present signal to the control valve with suitable transducers

Investment and Payback Period

The total investment required is Rs. 1,65,000, which will be paid back in 25 months.

5.2.10 Installation of Cyclic Load Energy Saver (Star-Delta Convertors)

Present Status

The efficiency of an induction motor varies with load. The efficiency drops steeply when the load at the motor falls below 50%. Hence it is desirable to run the lightly loaded motor in star mode instead of delta mode. When star mode is connected, the voltage applied across each phase of the motor and current are each reduced by a factor of 0.58. Hence the energy consumption of the motor is reduced by 15–20%. Keeping this in view, it is suggested that microprocessor-based energy savers working on the star-delta changeover principle may be applied to a very lightly loaded motor. After starting and running the motor in delta mode for a few seconds, this motor can be switched over to star mode and run continuously. When the loads are increased, it is automatically switched back to delta mode. In the same manner, these cyclic energy savers can be applied to the machines operating with variable loads, e.g., lathes, power processes. The expected power saving is around 25–30% of the no-load power drawn by the motor.

Star-delta starters are normally provided for SG motors with the rating above 20 HP. Nearly 80 motors exist in this range (above 20 HP). Most of these motors are functioning to their full capacity; only a few open with partial load.

Modification Proposed

On a trial basis, cyclic load energy savers may be provided for the following two motors:

1. Straight juice pump motor no. 2 = 30 HP
2. Air blower motor = 20 HP

Based on the success of this trial study, the application pf this energy saver for similar motor may be considered.

Investment and Payback Period

The estimated investment value is Rs. 80,640, and the payback period is 4 months.

5.2.11 Install Shunt Capacitors Directly across the Terminal of All Induction Motors in Steps, Particularly Motors Rated 50 HP and Above

Present Status

At present, shunt capacitors have been provided in the power distribution panel/motor control panel. For certain motors with a rating of 100 HP and above, shunt capacitors are provided at the motor terminals. The total VAR compensation in the plant is around 1350 KVAR, and the power factor is kept above 0.85 lag.

Modification Proposed

To get higher energy savings, it is always preferable to have the shunt capacitors connected either at or close to the motor terminals. Such a step will help to reduce the energy losses in the power delivery network, especially in the cables, as the power factor is corrected at the load end. It will help to achieve higher overall system efficiency than the method in which capacitors are provided on the upstream side only. In the instant case, it is suggested that these capacitors may be connected to all motor terminals in steps wherever they have not been provided earlier. This may be carried out depending on the site condition in stages. In other words, the practice already in existence in the plant may be applied on a wider scale.

Investment and Payback Period

The total investment value is Rs. 2,28,000, and the paid back period is 14 months.

5.2.12 Conversion of Lightly Loaded Delta-Connected Motor to Permanent Star Connection

Present Status

To achieve energy savings on a lightly loaded, three-phase, delta-connected SG induction motor, the voltage across the welding is decreased by permanently connecting the motor winding to star mode. When the motor is lightly loaded (less than 50% load), the starter connection to the motor may be permanently connected in star mode instead of the changeover to the delta mode. When the star mode is connected, the voltage applied across each motor phase is reduced by a factor of $1/\sqrt{3}$. The line current is also reduced by an improved factor. This results in a saving of 15–20% of the input power.

Modification Proposed

Hence it is suggested that it can be extended further by performing periodical motor performance studies.

Investment and Payback Period

The investment is nil, so "repayment" is immediate.

5.2.13 Periodical Motor Load Survey May Be Taken up in the Plant in a Methodical Way So as to Identify Further Improvement and Energy Saving Options

Present Status

Even though the plant has a massive population of LT motors of various capacities, a load survey has not been undertaken so far to identify improvement options.

Modification Proposed

The methodology as listed here may be adopted methodically. First is the selection of representative LT motor drives among the motor population. The criteria to be considered for this study are:

1. Utilization factor, i.e., hours of operation, with preference given to continuously operated drive motors.
2. Energy conservation potential basis, i.e., drive motors with inefficient capacity controls on the machine side, considering fluctuation load drive systems.
3. Measurements on the motors thus selected involve the electrical load parameters like voltage, current, power factor, power drawn, and energy consumption, as well as drive parameters like speed, load pressure, temperature, etc.
4. The main observations should indicate:
   - Percentage loading on the motor.
   - Percentage voltage imbalance, if any exists.
   - Quality of the power supplied.
   - Variations in voltage, current, frequency, and power factor.

5. Variations in machine-side load conditions like load/unload condition, pressure, flow, temperature, damper/throttle operation, whether it is a rewound motor, idle operations, meter provisions, etc.
6. End results/findings should indicate:
    • Identified motor with less than 50% loading, 50–75% loading, 75–100% loading, and exceeding 100% loading.
    • Identification in the motor of low voltage/power factor/voltage imbalance needed for improvement measure.
    • Identification in the motor of high machine losses/inefficiency, like idle operation, e.g., locations where the need for automatic controls/ interlocks or variable speed drives exist.

Energy Conservation Potential and Benefits

These kinds of load survey require no extra cost. They can be performed by the plant personnel themselves. They are aimed not only as a measure to identify motor efficiency but also as a means to check the combined efficiency of the motor-driven machines and controller. They can help to bring out savings in driven machines/systems, which can benefit from 30–40% energy savings in the course of time.

5.3 ENERGY CONSERVATION IN LIGHTING SYSTEM

5.3.1 Replace Incandescent Lamps and Blended Mercury Vapor Lamp with Compact Fluorescent Lamps

Present Status

The lighting conversion efficiency of the incandescent lamp is 13.8 lm/W (lumens per watt), which is very low. Blended mercury vapor lamps of 160 W installed have much higher luminous intensity than required. Blended lamps are very inefficient, and the lighting conversion efficiency is only 18 lm/W.

Modification Proposed

Replace incandescent and blended type mercury vapor lamps with compact fluorescent lamps (CFL).

Energy Conservation and Potential

1. Compact fluorescent lamps are much more efficient than blended and incandescent lamps. They give almost four times the light output of blended mercury vapor lamps and ten times that of incandescent lamps for the same power consumption.
2. The life expectancy of CFL is higher than that of blended and incandescent lamps.
3. The annual energy savings by replacing blended and incandescent lamps are Rs. 20,66,900.

Investment and Payback Period

Total investment for replacing incandescent lamps and blended lamps is Rs. 12,44,600, which will be paid back in 9 months.

5.3.2 Utilize Natural Light by Installing Translucent Sheets for Roofs in Plant

Present Status

Fluorescent lamps are used to illuminate 100 rooms even during daytime, since natural lighting is not sufficient. The plant has already installed translucent sheets in many offices and may install them in other offices in a phased manner.

Modification Proposed

It is recommended to install translucent sheets in the roofs to utilize natural lighting. After installing translucent sheet, lamps can be switched off for 8 hours a day.

Energy Conservation and Benefits

  a. There will be savings in lighting energy consumption since lamps can be switched off for 8 hours a day.
  b. The annual savings for installing translucent sheets on roofs in the plant is Rs. 21,480.

Investment and Payback Period

The total investment is Rs. 20,000, and the payback period is 7 months.

5.3.3 Install Photoelectric Control in Identified Areas to Control Artificial Lighting

Present Status

The lighting in the plant is mainly provided by fluorescent lamps. The shop areas are provided with a north light in the roof, which provides good lighting to the shop floor during daytime when the sky is clear. Apart from this, the machines are also provided with work lights. In spite of all these provisions, the shop's artificial lights are always switched on.

In many areas, the lighting from the nighttime lighting is found to be quite enough. The artificial lighting can be switched off as the intensity without them is sufficient, matching that of night values. The light and air circulators are connected in the lighting circuit without any segregation.

Modification Proposed

  1. Segregate the lighting and fan circuits, and provide distribution boards exclusively for lighting.
  2. Install photoelectric switches to switch off lights in identified areas.

Energy Conservation and Benefits

1. By switching off unnecessary lights during daytime with bright sun, 43,800 kWh of energy can be saved costing Rs. 1,57,000.
2. As kilowatt hours a day are reduced, lamp replacement frequency is also reduced.

Investment and Benefits

The estimated cost of segregating wiring and installing photoelectric sensor works out to Rs. 80,000 with payback in 6 months.

5.3.4 Install Solar Energy System, Install Solar Heater in Canteen/Guest House

Modification Proposed

1. **Solar heater:** It is suggested to install a solar water heater in the canteen. By installing these heaters, at least 8 months in a year, solar energy can be used. Existing heaters are to be retained for supplementing these units in case of bad weather or the rainy season.
2. **Solar lighting:** It is suggested to provide solar lighting for one of the street-lights and watch performance. If it is successful, it can be incorporated in other areas.

Energy Conservation and Potential

By installing a solar heater, there will be savings of Rs. 2,16,000 kWh costing Rs. 6,78,240.

Investment and Payback Period

The investment for the solar heater works out to be Rs. 4,53,750, and it will be paid back in 8 months.

5.3.5 Provide Reflectors in Each Fitting

Present Status

At present, no reflectors are provided in the light fittings.

Modification Proposed

To provide reflectors for all light fittings so that the illumination level of each fitting is increased. There will be a saving of 5% in the lighting load as the number of lamps can be reduced due to better illumination.

5.3.6 Installation of Energy Saver in Lighting Circuit

Present Status

The lighting to the plant is provided mainly by discharge lamps like blended mercury vapor lamps, sodium vapor lamps, and fluorescent lamps. In discharge lamps, the light output is roughly proportional to the input voltage. A reduction in voltage of about 5% does not cause a proportional reduction in light output. The light output is reduced marginally by 2%, but there is a substantial reduction of about 10% in power consumption. Similarly, a higher voltage does not give proportionally higher light output, but the power consumed is significantly high.

Though separate circuits for lighting with a light distribution board are available in some areas, air conditioners and fans are also connected to these lighting circuits. These are to be segregated into different circuits and not to be connected through energy savers. The total lighting load works out to 900 kW.

Modification Proposed

1. Segregate nonlighting loads like fans, air conditioners, and power plugs into separate circuits.
2. Balance the lighting load in each lighting distribution board as far as possible among the three phases.
3. Install an energy saver for each lighting distribution board.

The energy saver designed for the lighting circuit is a devices consisting of a negligible loss transformer, which, when introduced into lighting circuit, reduces the voltage input and consequently the power to the discharge lamps with a negligible reduction in light output. Nearly 25% of lighting energy consumed can be saved by installing energy savers. For estimation purposes, a very conservative figure of 10% is taken.

Energy Conservation Potential and Benefits

1. By installing these drives, the annual lighting energy consumption will be reduced by 7,68,960 kWh with a savings of Rs. 35,37,216.
2. The useful life expectancy and efficiency of the lamps also increases.

Investment and Payback Period

The estimated cost of energy savers and installation works out to Rs. 32,00,000/–, which will be paid back in 11 months on a simple payback basis.

5.3.7 Replace Electromagnetic Ballasts with Electronic Ballasts for Fluorescent Lamps

Present Status

Presently, fluorescent lamps operate with wire-wound ballast acting as a current-limiting element. These fluorescent lamps consume approximately 12 W of power more than that required with electronic ballast.

Modification Proposed

It is recommended to replace wire-wound ballast with electronic ballast.

Energy Conservation Potential and Benefits

1. Operates at an inherently high power factor; hence no external capacitor is required.
2. Instantaneous flicker-free start, and no starters are required.

3. With very low losses and low heat generation, the savings in power consumption is lessened.
4. Lamps operate under a wide voltage range of 120–270 V.
5. There is no optical illusion, ensuring safety.
6. The life of fluorescent lamps is enhanced.
7. The annual energy saving achieved by installing electronic ballast is approximately 38,000 kWh with a cost saving of Rs. 2,65,000.

Investment and Payback Period

Investment for electronic ballast is Rs. 4,00,000, and the payback period is 18 months. Electronic ballast is available with a number of manufacturers.

5.3.8 REPLACE STANDARD FLUORESCENT LAMPS WITH COMPACT FLUORESCENT LAMP

Present Status

Present lighting in the plant is provided by 36 W fluorescent tubes.

Modification Proposed

It is recommended to replace 400 of the standard fluorescent lamps by compact 11 W fluorescent lamps whenever they are changed out.

Energy Conservation Potential and Benefits

Compact fluorescent lamps are more energy efficient than 36 W fluorescent lamps. The energy saved will be Rs. 4 lakhs.

Investment and Payback Period

The total investment is Rs. 4,00,000, and the payback period is 4 months.

5.3.9 PROVIDE DAY LIGHT SWITCHES TO CONTROL LAMPS I N IDENTIFIED AREAS

Present Status

The process area of the plant is provided with enough lighting by means of fluorescent lamps. Fluorescent lamps are on throughout the day. It was observed that translucent sheets are not provided in the roof.

Modification Proposed

Install daylight switches to switch off lamps and provide translucent sheets in the roof to get natural light in daytime.

Energy Conservation Potential and Benefits

The saving in energy and cost by switching off the light in the identified area is 2,74,176 kWh costing Rs. 11,76,990.

Investment and Payback Period

The cost of daylight switches is Rs. 67,500 with a payback period of 1 month.

5.3.10 REPLACE SLIM TUBE FLUORESCENT LAMPS BY PHILIPS TRULITE TUBE LIGHT

Present Status

Presently, lighting in the plant is provided by 36 W fluorescent lamps.

Modification Proposed

It is recommended to replace slim tube fluorescent lamps with Philips Trulite tube light whenever the lamps are changed out.

Energy Conservation Potential and Benefits

Philips Trulite tube light is more efficient than slim tube 36 W fluorescent lamps. The lumen output from this lamp is 40% more than that of 36 W lamp but saves 10% of electrical energy. The energy saved will be 1,53,692 kWh costing Rs. 6,55,154.

Investment and Payback Period

As the slim tube fluorescent lamps will be replaced as and when they are changed, no investment is necessary.

5.3.11 CONVERSION OF HIGH-PRESSURE MERCURY VAPOR LAMP AND HALOGEN LAMP TO HIGH-PRESSURE SODIUM VAPOR LAMP

Present Status

At present, the high-pressure mercury vapor lamp of 250 W, 400 W capacity, high-pressure sodium vapor lamp of 250 W and halogen lamp of 50 W are used for street lighting.

Modification Proposed

1. The 250 W and 400 W high pressure mercury vapor lamps that are used for street lighting can be replaced with 70 W and 150 W high-pressure sodium vapor lamps.
2. The 250 W high-pressure sodium vapor lamp used for street lighting can be replaced with 70 W high-pressure sodium vapor lamps, respectively.
3. The 500 W halogen lamps used for street lighting and outside the factory can be replaced with 70 W high-pressure sodium vapor lamps.

Investment and Payback Period

Investment for the same lamps is Rs. 19,575, and the payback period is 3 months.

5.3.12 Install Lighting Transformers in Lighting Feeders

Present Status

There is no lighting transformer in lighting feeders.

Modification Proposed

It is suggested to install automatic servo voltage stabilizers or step-down lighting transformers in identified main lighting feeders and operate the lighting circuit at 205–210 V.

Energy Conservation Potential and Benefits

In discharge lighting, a 12% reduction in voltage results in:

1. A 12% reduction in power with negligible reduction lux levels.
2. A 1–2% reduction in illumination (negligible).

Investment and Payback Period

The investment value is Rs. 2.80 lakhs, and the payback period is 27 months.

5.3.13 Switch off Lights in Identified Areas

Present Status

In some sections of the plant, even though natural lighting was present, the light was on.

Modification Proposed

1. Install translucent sheets.
2. Regroup the lighting circuit and switch off unwanted lights.

Investment and Payback

Investment value is Rs. 20,000, and the payback period will be 21 months.

5.3.14 Replace Filament Indication Lamps in Control Panels with LED Lamps

Present Status

Filament lamps are used for on/off indication in control panels.

Modification Proposed

It is recommended to replace filament lamps with LED lamps, which consume 0.5–1.0 W per lamp. The filament lamp consumes 7–14 W per lamp.

Investment and Payback Period

The annual energy saving is Rs. 0.41 lakhs with an investment of Rs. 0.48 lakhs. The payback is 15 months.

5.3.15 Replace Machine Spot Lights with CFLS

Present Status

One hundred spotlights are attached to each machine during the production day. These spotlights are 40 W incandescent lamps.

Modification Proposed

It is suggested that spotlights be replaced with 9 W CFLs, which have the same light output and consume less power. These lamps may be connected to a motor circuit so that, whenever the motor is switched off, the lights are also put off.

Investment and Payback Period

Investment value is Rs. 1,12,939.

5.3.16 Replace Reflectors of Tube Light Fittings

Present Status

The fluorescent lights are used in the shop floor area. The power enameled reflectors are 5 years old and have lost luster.

Modification Proposed

Install anodized aluminum reflectors for all FTL fittings on the shop floor.

Energy Conservation Benefits

With the improved lux level, 10% of fluorescent lights can be switched off.

Investment and Payback Period

Investment value is Rs. 3,22,000, and the payback period is 26 months.

5.3.17 Balancing of Electrical Loads on Three-phase LT Lighting Feeder

Present Status

A separate three-phase low-tension (LT) circuit has been provided for the lighting load in the plant. On recording the current flows in the switchboard of this lighting circuit, it was found that the loads were not balanced, and hence current flowed through the neutral. Ideally, the neutral should not carry any current, but practically it is not possible to reduce it to zero.

Modification Proposed

Efforts can be made to reduce it by balancing the light loads. Higher current in the neutral will lead to energy losses, neutral overheating, and burnout.

5.3.18 Input Voltage Reduction to the Lighting Loads (400 FT Lamps + 60 Sodium Vapor Lamps of Various Capacities)

Present Status

At present the lighting loads are fed from a common distribution panel. No separate LT feeder has been provided for these lighting loads. The voltage input to these lighting loads is 230 V (in the range of 230–240 V) because the sugar plant is located very close to the cogeneration plant.

Modification Proposed

1. Install one automatic voltage stabilizer of capacity 50 KVA, for exclusively feeding of 400 FT lamps, 60 sodium vapor lamps of various power ratings, 50 CFLS, and 50 LEDs.
2. Provide separate three-phase LT circuit for feeding lighting load.
3. A marginal reduction in the input voltage to these lamp fittings will result in appreciable energy savings without affecting the lighting levels appreciably. Further, this step will help to reduce the failure rates of chokes, starters, tubes, and other accessories of the fitting. In the case of LED and CF lamps, the energy saving proposals as given in an earlier Energy Conservation Opportunity will be applicable.
4. A constant input voltage has to be maintained at 370 V if (L-L) and 210 V (phase–ground voltage for the lamp circuit) by suitably setting the automatic voltage regulator.
5. Adjust the upstream overcurrent protective relays in accordance with the operating power factor of this circuit since the input voltage to these fittings may lead to higher currents.

The expected energy savings as a result of reduction in input voltage is approximately 1% for every 1% reduction in supply voltage; i.e., 1% reduction in supply voltage saves nearly 1% of energy.

5.3.19 Replacement of 50 Existing FT Lamps with Latest LED Lamps

Present Status

Forty watt FT lamps are generally used in the office building.

Modification Proposed

It has already been proposed to replace 100 existing FT lamps in the residential quarters and streetlights with slim 36 W FT lamps and 50 FT lamp fittings in the office building with energy efficient CF lamps. Now it is proposed that 50 more FT lamp fittings in the office buildings and in the plant may be replaced with the modern LED lamps of 5.5 W.

Investment and Payback Period

Investment value is Rs. 75,000, and the payback period is 36 months.

6 Effect of EMI on Electrical and Electronic System and Mitigation Methods for Low- and High- Frequency Applications

Y. Uma Maheswari, A. Amudha,
and L. Ashok Kumar

CONTENTS

LEARNING OUTCOMES

At the end of this chapter, the reader will be able to understand:

- The fundamentals of electromagnetic induction (EMI) and elements of EMI.
- The effect of EMI on electrical systems and its impacts on energy.
- EMI mitigation techniques.
- EMI simulation and measurements.
- Energy efficient converter design topographies.

DOI: 10.1201/9781003203810-6

6.1 INTRODUCTION

Electromagnetic interference is the operational disturbance of electronic devices in the presence of electromagnetic waves caused by another device or source. Unwanted signal can be emitted by electrical or electronic circuits for various reasons. Electromagnetic emission happens due to high-speed clock signals, processor noise, high switching frequency, transmission interferences, or network interference. Due to this performance of the signal, the receiver circuit degrades. This causes the unintended operation or malfunction of electromechanical equipment, circuits, and components.

As shown in Figure 6.1, a source generates the interference, and it is conducted and radiated through a coupling path. Due to this interference, the receiver is not receiving the expected signal. Measures to be taken involve both the source and the transmission side or conduction path. Two of the most common sources of EMI are conversion of some differential signal into a common signal, which eventually gets out on an external-twisted pair cable, and ground bounce on a circuit board, generating common currents on external single-ended shielded cables. Additional noise can come from internally generated radiation leaking out of the enclosure.

The electromagnetic waves that cause interference in the electronic circuits are called electromagnetic noise. When an electronic device exposes electromagnetic waves, undesired electric currents can be induced in the circuit, causing malfunction of the circuit/system or the degradation in the performance, which is an unintended operation. The main events of this environment are source, conduction path, and victim where the noise interference happens. Based on source, noise can be classified into human-made and natural noise; because of either, noise is emitted. Noise is transmitted by conduction through wire and radiation through a nonconductive medium. Victims are the devices or appliances that receives the noise, and they need immunity.

On the source side, noise can be due to signal, power supply, and surge. The digital circuits become noise sources due to the transition of the signal having a wide range of frequency switching from 0 and 1 in a very short duration. Digital circuits transmit the data or information by switching the signal level between 0s and 1s to operate the circuits. While the signal levels are switched, high-frequency current flows into the circuit. The current flows in a signal trace, as well as power supply and ground planes. Power supply is also a cause of electromagnetic noise. Though the voltage is stable, the current varies according to the operation. A huge amount of high-frequency current flows inside the electric circuit. Since the power plane is shared among the PCB, the noise generated also gets transmitted to the entire circuit.

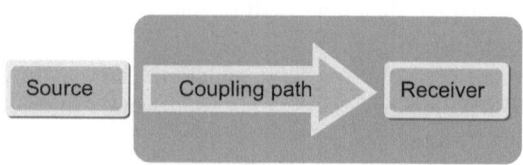

FIGURE 6.1   Elements of EMI.

Switching the power supply is also one of the noise sources due to the leakage current. Another source of noise is the sudden rise in the power supply, or surge, due to which there can be catastrophic or latent failures of the circuit.

Resonance is one of the factors for electromagnetic noise. Applications of damping resistors or ferrite beads help to suppress the noise. According to the transmission theory, the trace or conductor is considered a transmission line, wherein electric energy is expected to propagate and reflect as a wave. Tt can be either noise or signal and is considered wave propagation in a transmission path. When the digital signal is connected in a long trace and because the signal is reflected at both ends, it may cause a ringing effect. Hence the reflection needs to be suppressed by impedance matching of the conduction path. It is the so-called characteristic impedance of the traces matched with the impedance of the connected circuit.

When the circuit impedance is connected to the end of trace or terminals, which is similar to the characteristic impedance, it is said that there is no reflection and that 100% of the electric energy is transmitted to the load. In this case, the terminal is considered to be impedance-matched. The load has received all the energy transferred without any reflection only when the characteristic and load impedances of the transmission line are matched. Implementation of impedance matching mitigates noise conduction. The S-parameter (scattering parameter) calculation method helps to measure the circuit characteristics during high-frequency operation. The S-parameter results in changes with the frequency. The S-parameter of a device has to be set for the input and output ports. The result of simulation is in terms of the transmission coefficient and reflection coefficient, which is gotten by sending an electric signal through the ports.

The fundamental properties of the signals and their interaction with interconnects are determined with the help of time domain and frequency domain. Looking into various perspectives of a signal sent through an interconnect is called a domain. The time domain is considered a reality, and the digital signals can be analyzed as we can measure the performance extensively. In this domain are two main terminologies of clock signal: clock period and rise time. Frequency domain is a mathematical representation. Impedance can be perceived in both the time and the frequency domains. The impedance concept can be easily understood in the frequency domain. Fourier transform functions are more helpful in converting time domain to frequency domain. Three main types of Fourier transforms are Fourier integral (FI), discrete Fourier transform (DFT), and fast Fourier transform (FFT). Each can be applied in different conditions. A series of discrete points in the sine wave can be converted to frequency domain where the time period of the wave may not be same for all of the cycle. In the case of a discrete Fourier transform, the waveform is regular, periodic, and repeating at a regular time period. Fast Fourier transform is similar to discrete Fourier transform but uses a fast algorithm for the calculation. Inverse Fourier transform can be applied to convert from the frequency domain to time domain. Here each frequency component is taken and converted it into its time domain sine wave, and then it is added to the rest of the frequency component. The main reason for normally using the frequency domain is the fastness compared to time domain results.

There are various techniques to suppress or mitigate the noise levels generated from electronic devices for the circuit to perform as intended. Filters are used to remove unwanted electric current flowing in conductors and divert it to ground. Either the noise can be transmitted inside the parts, or it is returned to the originating noise source. Application of a shield is used to suppress the noise conducted through the radiation model. Noise suppression usually applies low-pass filters that remove high-frequency components. Passive components such as inductors, resistors, and capacitors can be used for the low-pass filter. In addition, optical cables and transformers can also be used.

6.2 EFFECTS OF NOISE

Electromagnetic waves are the unwanted signals that can be present in radio waves, infrared, microwaves, X-rays, ultraviolet, light, and gamma rays. Due to electromagnetic radiation, electronic circuits may malfunction, or there can be degradation in performance. In addition, the electromagnetic field (EMF) generated from various applications, such as telecommunication, electrical appliances, medical equipment, and other apparatus, which involves digital and high-speed signals, can definitely impact human health. Prolonged exposure to an electromagnetic field may result in a higher risk of health issues like cancer, brain tumors, neurological, and physiological problems by disrupting human nerve function and other health problems.

6.3 STANDARDS FOR EMI MITIGATION

EMI mitigation is required for high-density semiconductor packaging technology advancement with complex stacked chip and multichip packages. This is equally important due to the other reasons such as the increase of electromagnetic pollution and emergence of the Internet of Things (IoT). EMI is regulated and enforced per international and national standards in order to overcome these problems. Some of the regulatory bodies are:

- International Electrotechnical Commission (IEC).
- Federal Communications Commission (FCC).
- Verband Deutscher Electrotechniker (VDE).
- International Special Committee on Radio Interference (CISPR).
- IEEE (Institute of Electrical and Electronics Engineers).

Other standards organizations, including ISO, SAE, also define a number of standards geared toward specific applications or industries.

Based on these standards, the product is tested in the test laboratory using various instruments, such as the spectrum analyzer, antenna, signal generator, ESD generator, oscilloscope, EMI test receiver, surge generator, etc. All of these instruments are designed according to a defined basic standard. Using the report from the test laboratory, the product can be applied for regulation. Different countries follow different regulations, such as FCC, CE, CCC, etc.

Standards are classified as basic standard, generic standard, product-oriented standard, and product-family-oriented standard. They are based on the frequency

TABLE 6.1
Applications and Their Frequency Range.

Description	Frequency range
Wireless	500 MHz+
Wi-Fi	2.4–5 GHz
Bluetooth	2.4–2.485 GHz
4G cellular	0.7–2.7 GHz
5G cellular	6 GHz

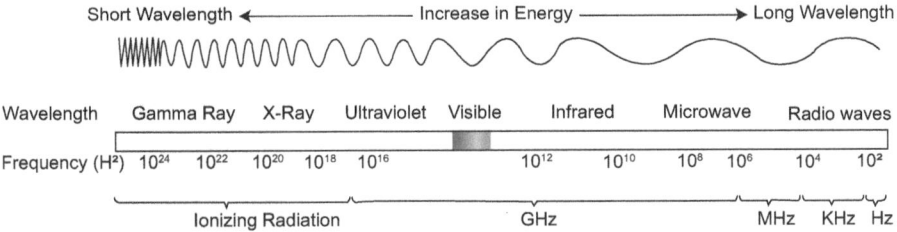

FIGURE 6.2 Frequency ranges with wavelengths.

band or wavelength and the testing parameters or metrics. Figure 6.2 shows, in general, the various frequency ranges with wave lengths.

With the different frequency ranges, there are various applications. The Table 6.1 shows the typical communication examples with their operating frequency.

The EMI problem arises at high frequencies, and it has to be watched at various levels to be overcome.

6.4 EMI AT COMPONENT LEVEL

Behavior of an electronic component in low frequency differs from what can be expected at high frequency. For example, a resistor under low frequency works as a resistor as expected. But during high-frequency operation, the equivalent circuit will not be just a resistor; instead, some inductance effect is added. This is the same for other components. Figure 6.3 shows passive components and their behavior at high and low frequencies. In a resistor, at higher frequencies, however, parasitic capacitance (between windings or end-to-end) limits impedance. Capacitor impedance is actually low at resonance, which might be seen as good. But above the resonant frequency, the capacitor actually behaves like an inductor, increasing in impedance with frequency.

The selection of capacitor is important based on the operating frequency. With the lead of the capacitor, the performance will degrade. EMI shielding is applicable to various component levels, such as system-in-package (SiP), system-on-chip (SoC), microcontrollers, application processors, power amplifiers, wireless modules, radio frequency modules, memory, sensors, digital signal processors (DSP), application-specific integrated circuits (ASIC), field programmable gate arrays (FPGA), analog digital converters (ADC), and so on.

6.5 EMI AT PCB LEVEL

In a printed circuit board (PCB), emission and radiated emission is conducted due to the power supply and signal speed. The source of the EMI is the place or module where the interference is generated. A receptor for EMI is the block being affected by the interference. PCB has various modules such as power supply, memory, filters, synchronizer, and other components—cable, PCB traces, connector, res-ind-cap, switch. Each module has to be monitored for EMI. Two different domains are used to analyze EMI: the time and frequency domains. In the time domain, signals are analyzed with respect to time. In the frequency domain, signals are analyzed with respect to frequencies. In the very-high-frequency signal, PCB requires an electronic band gap (EBG), which acts as a low-pass filter, which is used for mitigating EMI issues.

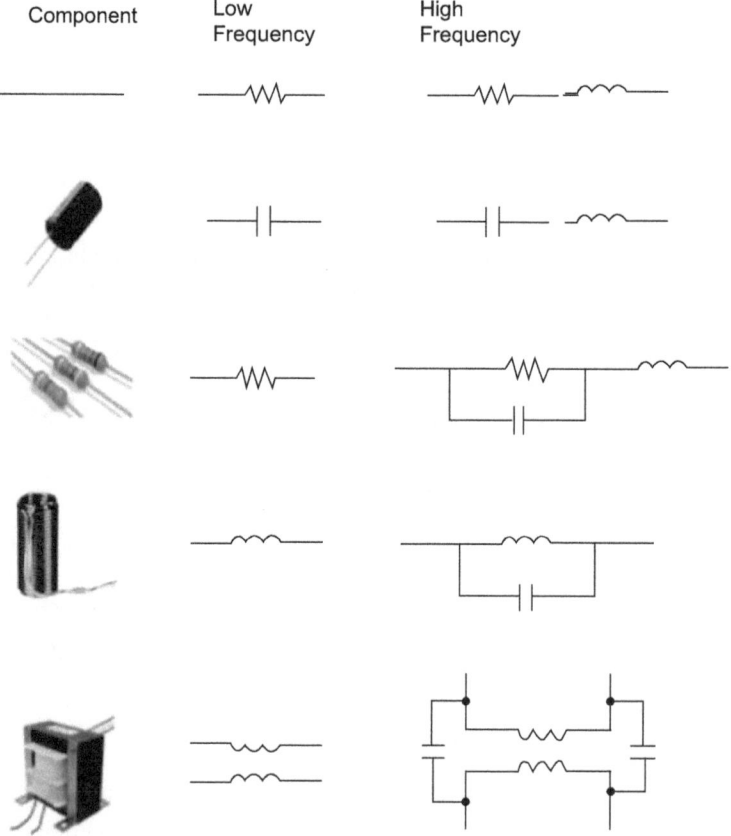

FIGURE 6.3 Equivalent circuit of component with operation frequency.

TABLE 6.2

Applications and Their Frequency Range.

Type	Maximum frequency
Aluminum electrolytic	100 kHz
Tantalum electrolytic	1 MHz
Paper	5 MHz
Mylar	10 MHz
Polystyrene	500 MHz
Mica	500 MHz
Ceramic	1 GHz

In the multiple-layer high-speed PCB design, on the plane layers (power and ground), the ground bounce noise (GBN), also called simultaneous switching noise (SSN), is becoming the challenging factor with system-on-package, which is the integration of high-speed microprocessors, radio frequency (RF) circuits, memory, sensors, optical devices, etc. With the increase in higher system operating frequencies, resonance can happen due to the plane layers, and this acts as a radial wave that travel outward to the edge of PCB from the noise source. Some of it is reflected inward, creating signal integrity (SI) issues, and some part is radiated into a nonconductive path, creating electromagnetic interference (EMI) issues. In the case of ultrawideband (UWB) communication technology, the signal power is very minimum, and reflected noise from the PCB will affect the functionality of the RF circuits.

In order to avoid the noise, some of the basic techniques such as the addition of a decoupling capacitor, plane layer segmentation and an isolation moat can be applied. But these are effective at low frequency and not suitable for frequency ranges above 1 GHz. Another approach to eliminate such noise is the application of electromagnetic band gap (EBG) structures. Much research is going on in EBG structure design for the suppression of GBN.

In a printed circuit board, copper traces or pattern is considered a transmission line. This transmission line behaves differently based on the frequency of the signal that propagates through it. Also, it depends on the dielectric constant of the board material. The transmission line characteristic impedance is the instantaneous impedance of a signal measured when it propagates forward. The characteristic impedance is not relevant with the trace length but is related to the signal speed and capacitance per unit length. Controlled impedance of the entire board has to be maintained for all traces, failing which impedance discontinuity can occur and lead to signal integrity issues in the case of high-speed design applications. Because of impedance discontinuity, reflections and distortions occur, affecting signal quality. Reflection is referred to as the portion of the signal reflected back to the origin or source.

In order to get a good-quality signal, it is important to control impedance for interconnects and termination, adopt proper routing topology, avoid multiple branches, and minimize geometric discontinuities. Other parameters or factors are stub, neck down races, via between signals, package lead, input gate capacitance affects, and the impedance of the circuit.

Irrespective of the domain (time or frequency), the S-parameter is used in electromagnetic simulations where the characteristics of an interconnect can be studied, including impedance details, cross talk details, and attenuation details. S-parameters describe how the signal is propagated. For measuring the S-parameter, the port is set, and the signal to be simulated is sent through the DUT (device under test), and this has the signal and return paths. In order to track the sine wave entry and exit, each port is named with index numbers.

The S-parameter is the ratio between the output and input sine waves. To identify the input and output ports, two indices are adopted. Figure 6.4 depicts the S-parameter model of two ports (input and output) where 1 represents the input port, and 2 represents the output port. The signal enters through the input port and exits through the output port. If there is no loss condition, the entire signal entering into the device is transmitted to the receiver side. In the case of loss or impedance discontinuity, only a partial signal is transmitted, and the rest of the signal is reflected. S11 and S22 represent the reflection coefficient, and S21 and S12 represent the transmission coefficient. Based on the S-parameter value measurement, it is easy to conclude the electromagnetic emission.

A signal entering into port 1 and exiting in port 2 is labeled S21. Similarly, a signal entering into port 2 and exiting in port 1 is labeled S12. In the case of a dual port device, there can be three unique S-parameter values: S21, S11, and S22. The magnitude of the reflected S-parameter S11 is termed return loss, S22 is the insertion loss, and S21 is the insertion loss. Here S21 and S12 are unique. In a two-port network, S11 and S22 contain the information about the impedance discontinuity, and S21 and S12 contain the information related to losses and coupling with other lines. The standard impedance of a port is normally set as 50 Ω.

Various measures and solutions in hardware or software, as well as at system level, can be taken in order to deal with this issue. One such solution is implementation of an electromagnetic band gap (EBG) pattern in the PCB. Isolation in the plane layer provides the better noise reduction in radio frequency and analog circuits. It is essential to develop such filters for all kind of frequency applications. To

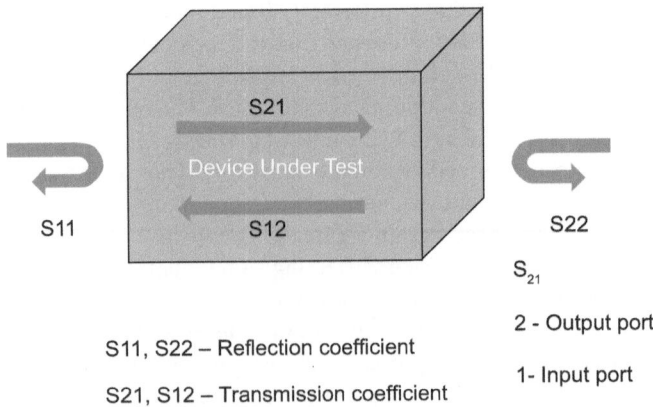

S11, S22 – Reflection coefficient

S21, S12 – Transmission coefficient

S_{21}

2 - Output port

1 - Input port

FIGURE 6.4 S-parameter model.

suppress electromagnetic noise, an electromagnetic band gap technique has been implemented in the power distribution network to obtain higher impedance with good SSN noise attenuation. For this reason, many EBG structures are proposed and applied. Of these, the mushroom structure with and without via and multilayers was studied. EBG is considered a periodic or regular geometric structure, and it acts as a high-impedance surface to an electromagnetic wave in a particular range of frequency. It can be designed as microstrip or strip line structures with rectangular patches implemented on a PCB. EBG filters can perform well at greater frequencies compared to passive filter components and ferrite counterparts. EBG can be implemented by integrating it in PCB layout with or without additional via beneath the signal path. EBG structures can be applied in plane layers such as power and ground planes, and partially placed EBG have also been taken for research recently.

Electromagnetic band gap (EBG) structures are also known by various names: photonic band gap (PBG), high-impedance surface (HIS), metamaterials (MTMs), or electromagnetic crystals (ECs). An EBG is applied in order to avoid electromagnetic wave propagation, and it exhibits frequency band across the structure. Electromagnetic band gap structures are patterned structures within which the unit cell size is smaller than the free space wave length. The unit cell is periodically step-and-repeated and is the EBG structure. This EBG structure, due to the suppressing property (of the surface wave propagation) in a particular frequency band, can be employed in various applications. The pattern can be embedded in the ground plane of the PCB. Some of the examples are microwave filter design, packaging technology, printed circuit boards, antenna technology, etc.

Primarily two types of EBG structures are available based on the position of the unit cell: mushroom type and uniplanner structure. The mushroom-type structure is a very basic structure where the copper patches are connected to the ground plane through vias. The mushroom-type EBG structure is smaller in size than the uniplanar type and has a lower frequency of operation. In the uniplanar EBG surface contains a structure where the patches or unit cell is directly connected with the ground plane without vias. It offers more advantages than the mushroom type when it comes to performance. The unit cell can be modified with interconnected sections of meander lines, and the uniplanar compact photonic band gap (UC-PBG) structure is achieved. These lines are considered the inductance and capacitance effect, which forms an LC circuit.

6.6 EMI MITIGATION TECHNIQUES

There are three types of solution for EMI issues:

1. Hardware solution
2. Software solution
3. System-level solution

USING HARDWARE

Fundamental and lower-order harmonics produce the radiation. This radiation interferes with other possible signals. Different reduction or elimination techniques can work for electromagnetic interference issues by applying the following:

- Fabric-over-foam (FoF)
- Shielding:
  - Coaxial transmission line method
  - Dual TEM cell method
  - Rectangular waveguide method
  - Nested reverberation chamber method
  - Shielded box method
  - Shielded room method
  - Free-space method
- Conductive filler CM and DM noise
- Form-in-place
- Ferrite
- Grounding
- Bonding,
- Spatial placement of hardware
- Circuit topology modification and spread spectrum

Noise due to high frequency in power lines can be eliminated by adopting the proper filters in the circuit. EMI filters normally consist of a combination of inductors and capacitors. The requirement for capacitors and inductors is based on node impedance where capacitors are used in high-impedance nodes and inductors are used in low-impedance nodes. Enclosing the system with a conductive material completely, called shielding, eliminates the EMI effect. This is the costliest solution, but it works. Radio-frequency- (RF-) emitting devices have to be isolated to limit the propagation of their interference to nearby components. Nowadays, the world is moving toward miniaturization, light weight, and high speeds. This becomes a challenge in overcoming the problem.

Using Software

Various software tools are available in the market to test EMI parameters in the printed circuit board. Once the native CAD design files are imported into the tool, the design rules can be checked. Prior to the simulation, the EMC engineer identifies the critical nets and critical components that are the most important to EMC performance. Clock signals, fast rise time, or data rate data signals are called critical nets. Critical components might be decoupling capacitors, filter components, termination resistors, etc.

System-level Solution

Optimization of the assembly-level code improves the EMI-related problem. This method eliminates the physical redesign of the system.

6.7 EMI SIMULATION

During the design stage of EMI analysis, S-parameter (scattering parameter) model extraction in the frequency domain can be done to know transmission and reflection characteristics by finding out the amplitude and phase information of a signal. Three types of simulators are available that can be performed to find the EMI levels before manufacturing the product.

1. **Electromagnetic (EM) simulators:** These solve Maxwell's equation and simulate the electric and magnetic fields at various locations in the time domain or frequency domain. Several EM simulation technologies have emerged based on Maxwell's equation. Electromagnetic full-wave simulation includes the method of moments (MoM), finite element (FEM), finite difference time domain (FDTD), and the partial-element equivalent circuit (PEEC) technologies.

2. **Circuit simulators:** These solve the differential equations corresponding to various circuit elements and include Kirchhoff's current and voltage relationships to predict the voltages and currents at various circuit nodes, in the time or frequency domains. It is a mathematical model depicting the behavior of an actual circuit.

3. **Behavioral simulators:** These use models based on tables and transmission lines and other passive element models based on transfer functions, which quickly predict the voltages and currents at various nodes, typically in the time domain.

6.8 DC–DC CONVERTER DESIGN AND TOPOLOGY SELECTION

In a modern automobile, the most commonly used power converter is the DC–DC converter. These converters are mainly used to power all the electronic loads present in it. Loads can be the engine control unit (ECU), infotainment system, etc. In modern times, with the advent of electric vehicles (EVs), hybrid electric vehicles (HEVs), and plug-in hybrid electric vehicles (PHEVs), and with better battery material technology, the nominal voltage of batteries tends to increase. One modern battery voltage that is widely becoming mainstream is the 48 V battery voltage. This nominal battery voltage is highly preferred, as it is more efficient in driving permanent magnet synchronous motors (PMSM), commonly used in electric traction, at this voltage than at 12 V. Therefore, it also becomes widely important to develop power converters that can power the electronic systems at the required voltage level from a 48 V power source. Therefore, a 48 V/12 V DC–DC converter is developed that can be used to convert the 48 V nominal battery voltage to 12 V, which can be further reduced to the required levels by the existing DC–DC converters. All electronic systems in an automobile have to undergo stringent conducted and radiated EMI emission standards. Therefore, the power converter design will also incorporate EMI filters to reduce conducted EMI emissions.

For implementing a 48/12 V DC–DC converter, various topologies are available. For powering electronic systems, bidirectional converters are not required. Likewise, isolated topologies are also not preferred due to higher cost, volume, and weight of isolating transformers, which have leakage inductance that contributes to conducted and radiated EMI noise. Therefore, the nonisolated buck topology is selected to implement the 48/12 V DC–DC converter.

TABLE 6.3

Converter Specifications.

Input voltage	48 V
Output voltage (Vo)	12 V
Rated power	100 W
Output voltage ripple	5% of V_o
Output current ripple	5% of I_r
Rated output current (I_r)	8.333 A

6.8.1 CONVERTER SPECIFICATIONS

The specifications of the 48/12 V DC–DC converter are shown in Table 6.3
    The design of a 48/12V nonisolated DC–DC buck converter is as follows:

- Duty cycle: $V_o/V_{in} = 12/48 = 0.25$
- Rated power: $P_{rated} = 100$ W
- Rated output current: $I_r = P_r/V_o = 100/12 = 8.33$ A
- Inductor current ripple: $\Delta IL = 5\%$ of $I_r = 0.4167$ A
- Output voltage ripple: $\Delta V_o = 5\%$ of $V_o = 0.6$ V
- Switching frequency: $f_s = 166.7$ kHz ($T_s = 6$ µs)

Therefore, the values of the inductance and capacitance are as follows:

$$L = \frac{V_0*(1-D)}{\Delta IL*f_s} = \frac{12*(1-0.25)*6*10^{-6}}{0.4167} = 129.6 \ \mu H$$

$$C = \frac{\Delta IL}{8*\Delta V_0*f_s} = \frac{0.4167*6*10^{-6}}{8*0.6} = 520.9 \ nF$$

Figure 6.5 shows the schematic diagram of a DC–DC converter circuit. The LISN (line impedance stabilization network) symbol is inserted to measure the conducted EMI as per the standard.
    Also for measuring conducted EMI, the components used in the power converter design have to be considered for nonideal parasitic components, both active and passive. Conducted EMI noise tends to peak at the self-resonating frequency due to low impedance for the resonating frequency component of the current signal. Therefore, it becomes very important to include all the nonideal characteristics of active devices as well as the parasitic components of both active and passive devices before measuring conducted EMI.
    Figure 6.6 and Table 6.4 shows the conducted EMI measurements without filters.

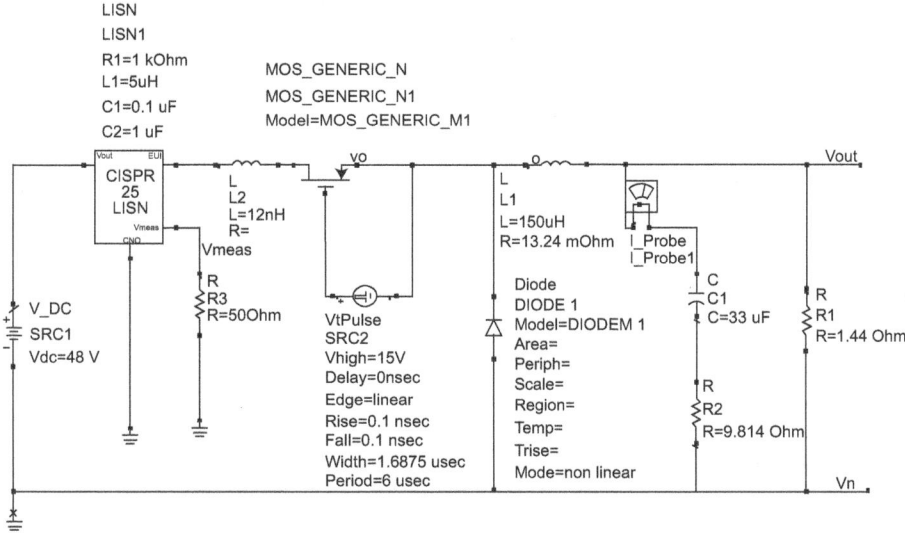

FIGURE 6.5 Schematic of converter circuit.

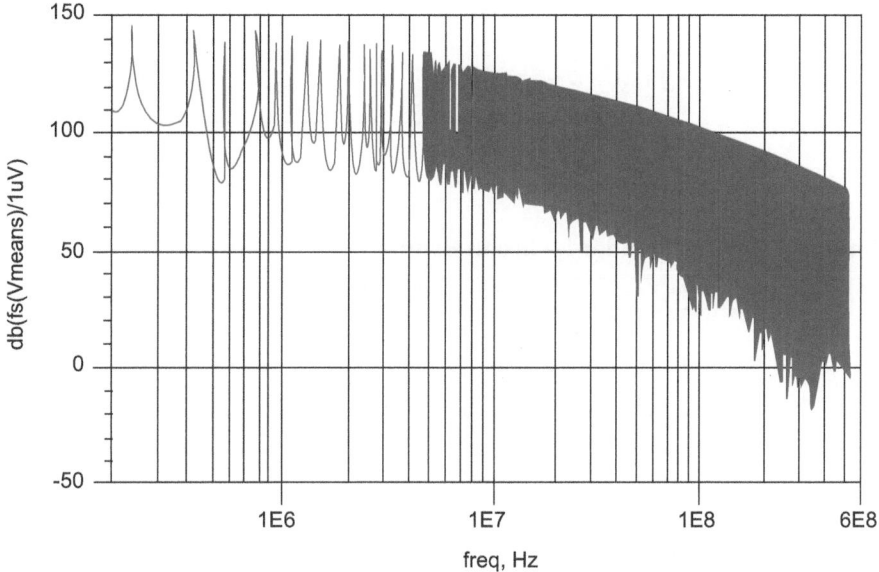

FIGURE 6.6 Measurement of conducted EMI.

6.8.2 CISPR 25

The most commonly used EMI measurement and conformance for the automotive application standard is CISPR 25 [5]. This standard defines the controlling of electromagnetic interference in electrical and electronic devices and is a part of the International Electrotechnical Commission (IEC). CISPR 25 details the limits and procedures for the measurement of onboard radio disturbances in the range of 150 kHz to 2500 MHz. In the CISPR 25 standard, the conducted EMI noises are classified into narrowband and broadband noise.

Each of these noises has its own limits as specified in the CISPR 25 standard. For these limits, the broadband noises can be measured by using an average or quasi-peak detector, and the narrowband noises can be measured using a peak detector. Tables 6.5 and 6.6 show the broadband and narrowband limits respectively.

It is evident that the designed power converter does not confirm to CISPR 25 standards from Table 6.4. Therefore, in order to satisfy the standard, an EMI filter (low pass) must be included in the design to reduce conducted EMI emissions.

TABLE 6.4
Measurement of Conducted EMI Emissions without Filter.

Frequency (MHz)	Design values (dBμV)
0.15–0.3	143
0.53–2	142
5.9–6.2	126
30–54	114
70–108	104

TABLE 6.5
Limit of CISPR 25 Standard—Broad Band.

Class	Levels in dB (μV/m)									
	0.15-0.3 MHz		0.53-2.0 MHz		5.9-6.2 MHz		30-54 MHz		70-108 MHz 144-172 MHz 420-512 MHz 820-960 MHz	
	P[1]	QP[2]	P	QP	P	QP	P	QP	P	QP
1	96	83	83	70	60	47	60	47	49	36
2	86	73	75	62	54	41	54	41	43	30
3	76	63	67	54	48	35	48	35	37	24
4	66	53	59	46	42	29	42	29	31	18
5	56	43	51	38	36	23	36	23	25	12

1) Peak
2) Quasi-peak

NOTES
1   For short duration disturbances, add 6 dB to the level shown in the table.
2   All values listed in this table are valid for the bandwidths specified in table 3.

TABLE 6.6

Limit of CISPR 25 Standard—Narrow Band.

Class	Levels in dB (μ V/m)				
	0.15-0.3 MHz	0.53-2.0 MHz	5.9-6.2 MHz	30-54 MHz	70 - 108 MHz 144 - 172 MHz 420 - 512 MHz 820 - 960 MHz
1	61	50	46	46	36
2	51	42	40	40	30
3	41	34	34	34	24
4	31	26	28	28	18
5	21	18	22	22	12
NOTE - For 87 MHz to 108 MHz, add 6 dB to the level shown in the table.					

6.9 EMI SIMULATION AND MEASUREMENT WITH FILTER

EMI filters can be designed using various topologies [3]. Some of the most commonly used topologies are π-type filter, L-type filter, T-type filter, dissipative filter, Cauer filter, and RC shunt filter. Some of the different filter topologies cannot be used in this power converter design due to high output impedance at the switching frequency, high component count, and standalone capability. It is possible to use a π-type filter or L-type filter with single-level or multilevel printed circuits.

6.9.1 EMI Filter Requirements

These conducted EMI emissions must be reduced from the values in Table 6.7 to the values in Table 6.6 corresponding to the required class of power converter. To achieve this, the attenuation requirements of the EMI filter for each class of CISPR 25 standard are as shown in Table 6.6. These values are calculated by computing the difference between the conducted EMI emissions obtained by simulations and the narrowband CISPR 25 EMI emissions limit corresponding to each class. The values are further decreased to ensure that the emissions with filters are well below the limits specified by CISPR 25 standard.

For a given low-pass filter, if the attenuation at frequency f_1 is $A1$ and at frequency f_2 is $A2$, then the order of the filter is:

$$n = \frac{A1 - A2}{6 * \log_2(\frac{f_2}{f_1})} \qquad (6.1)$$

For filter design, it is confined for class 4 and:

- $A1 \rightarrow$ Required attenuation at $f_1 = 0.15$ MHz
- $A2 \rightarrow$ Required attenuation at $f_2 = 0.53$ MHz

TABLE 6.7

EMI Attenuation.

Frequency (MHz)	0.15–0.30	0.53–2	5.9–6.2	30–54	70–108
Class	Attenuation (dB)				
1	−55	−80	−70	−65	−65
2	−65	−85	−80	−70	−70
3	−75	−95	−85	−75	−75
4	−85	−105	−90	−85	−85
5	−95	−110	−95	−90	−90

The order of the filter must be greater than or equal to the value of n calculated using the formula.

For a given nth-order filter, if the attenuation at frequency f_2 is $A2$, then the cutoff frequency is as follows:

$$f\text{cutoff} = f_2 * 2^{\left(\frac{A2}{6n}\right)} \qquad (6.2)$$

The preceding expression is used to determine the cutoff frequency of the required nth-order filter by substituting $A2$ as required attenuation at $f_2 = 0.15$ MHz.

The cutoff frequency of the filter is in terms of the inductance and capacitance is as follows:

$$f\text{cutoff} = \frac{1}{\pi\sqrt{LC}} \qquad (6.3)$$

By taking the standard values of capacitance C, the inductance L is determined, and the EMI filter is defined.

Class 4 Design

For a Class 4 design, the order of the filter must be greater than n; hence n is calculated using Eq. (6.1):

$$n \geq \frac{A1 - A2}{6 * \log_2(\frac{f_2}{f_1})} = \frac{(-85) - (-105)}{6 * \log_2(\frac{0.53}{0.15})} = 1.830$$

Therefore, for a Class 4 design, a second-order filter can be used, but such a filter will have a very low cutoff frequency and thus cause the output voltage waveform to ripple beyond the specified limits. Therefore, a third-order filter is used, and the cutoff frequency from Eq. (6.2) is:

$$f\text{cutoff} = f_2 * 2^{\left(\frac{A2}{6n}\right)} = 0.15M * 2^{\left(\frac{-85}{6*3}\right)} \cong 5KHz$$

For a third-order filter, an π-type filter is used.

Therefore, Capacitor value = $C/2$ = 100 μF (standard value) C = 200 μ and substituting in Eq. (6.3):

$$L = \frac{1}{\pi^2 * f^2 * C} = \frac{1}{\pi^2 * 5000^2 * \left(200 * 10^{-6}\right)} = 20.26 \, \mu H$$

Since the inductor is split between the power and neutral line:

$$\text{Inductor value} = \frac{20.26}{2} = 10.13 \, \mu H$$

The calculated filter components are incorporated in the converter circuit. Figure 6.7 shows the schematic circuit with filter components. The conducted emission is measured with a filter circuit and plotted.

Figure 6.8 shows the measurement of output voltage and conducted emission; it is also observed that the emission levels are well satisfied with the CISPR 25 standard.

Noise currents can be circulated internally by proper design of the circuit layout. The common mode current can be limited to within the power converter by ensuring that the CM noise generated by the MOSFET switching is absorbed by the parasitic capacitors in the circuit acting as Y-capacitors. A proper design of the PCB helps to adjust the values of the filter's capacitors. Wind band gap (WBG) devices make the converters more efficient and smaller in size compared to the silicon MOSFETs. The same converter circuit as used with silicon MOSFET is modified to suit the SiC and GaN MOSFET data sheet parameters. Both were simulated with Keysight ADS for conducted EMI noise. Both the SiC and GaN devices have significantly reduced on-state resistance as compared to the Si counterpart but have similar EMI noise.

6.10 ELECTROMAGNETIC BAND GAP STRUCTURE

Electromagnetic band-gap (EBG) structure is a periodic geometry on a dielectric substrate that acts as a stop band. This is used for blocking the electromagnetic wave propagation of certain frequency bands. EBG application has certain properties, where for a certain frequency range it blocks the surface wave propagation, and, without any change in phase, it reflects back any incoming wave. This can be applied in high-speed PCBs where it is needed to suppress electromagnetic noise.

Basically, the EBG structure reflects the electromagnetic noise of the frequency bands it can detect. Efficiency of the structure can depend on the patches at a higher density.

In addition to this band gap feature, EBG also possesses high impedance property and artificial magnetic conductor (AMC). The major issue faced in multiple-layer printed circuit boards at higher frequency operation is simultaneous switching noise (SSN) leading to SI (signal integrity) and EMI (electromagnetic interference), which

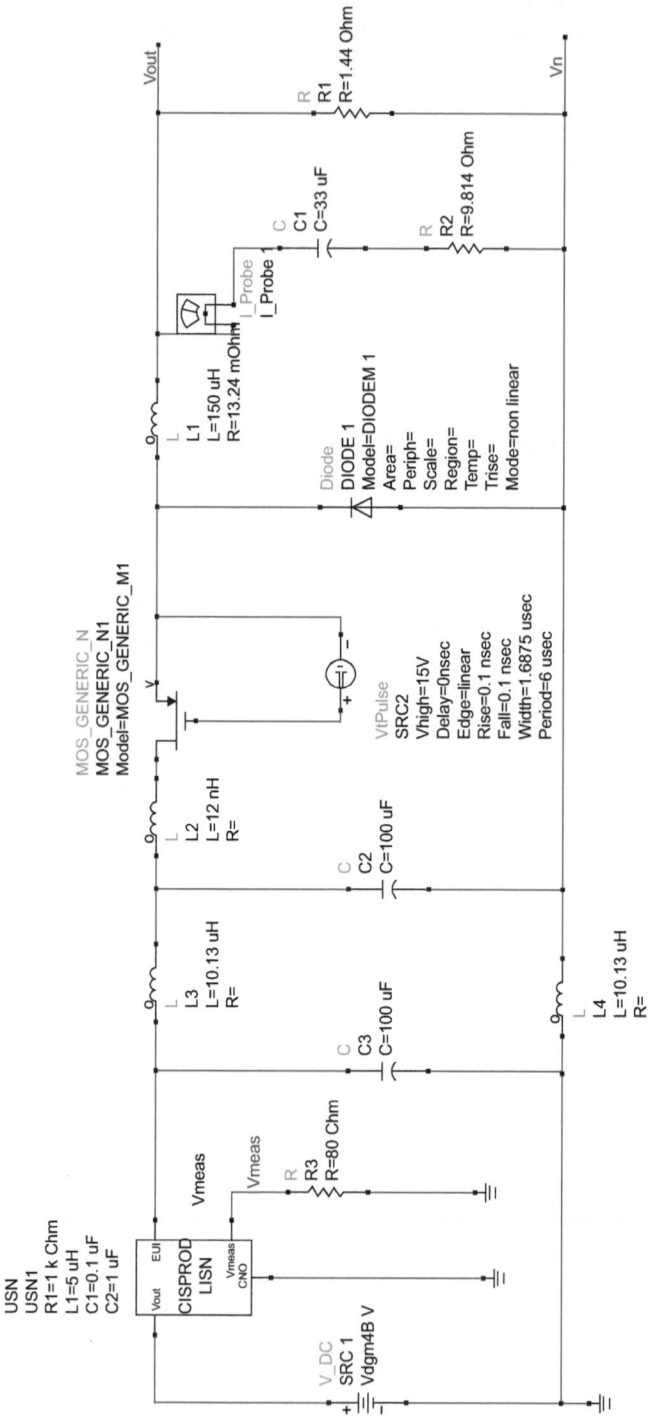

FIGURE 6.7 Schematic of converter circuit with filter.

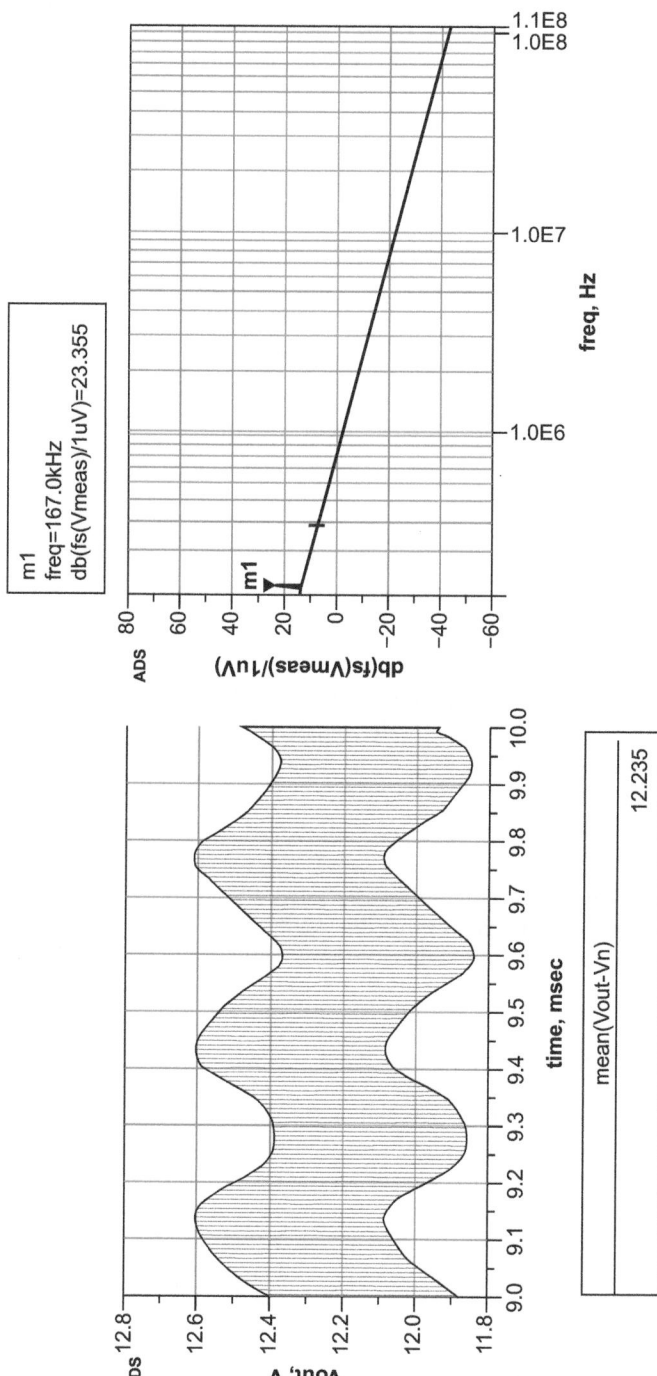

FIGURE 6.8 Measurement of output voltage and conducted emission.

can be minimized by the application of EBG structure. It can be implemented as a superstrate or as a ground plane in a PCB.

If the unit size is increased, capacitance value is increased, and consequently the stop band shifting happens at the lower frequency. If the gap in the unit elements is increased, capacitance value is increased, and the stop band will shift along a higher frequency range. If the via diameter is increased, the via impedance becomes smooth, and, due to this, the stop band shifting occurs toward the higher frequency side. A stop band is a band of frequencies within the specified limits; a filter circuit does not allow signals to pass through.

EBG structure can be classified broadly into three groups based on the band of frequency it operates in:

1. Single band
2. Dual band
3. Multiband

Various EBG structures exist:

- Mushroom type
- Crosshair
- Swastika type
- Hexagonal patch type
- Hexagonal patch with double slot type
- Square-patch-type with a single disconnected loop-type slot
- Fractal EBG type, etc.

In this chapter, a planar EBG structure with double-layer PCB is used for mitigating the noise level. Planar-type EBG structures has advantages in the multiple-layer printed circuit board with or without the presence of vias, and attention needs to be paid to electromagnetic leakage through the perforated layer, which can lead to radiation increase from the PCB. This chapter introduces the application planar EBG power plane as a shield to mitigate EMI, a novel concept of application in power planes.

6.11 EBG STRUCTURE DESIGN

In this paper, the electronic design automation (EDA) tool from Keysight Technologies, called Advanced Design System (ADS), is used for simulation in order to model the EBG structure and to measure the conducted EMI noise. Then the noise data are used to design filters, which are also incorporated into layout and simulated to verify their effectiveness. Keysight ADS is a software package that facilitates each and every step of the design process. The process flow in the tool starts from schematic capture: here it is possible to draw the equivalent circuit of the design layout; to draw the actual copper pattern, DRC-design rule checking; and to measure violations of the rule set in the software, frequency domain and time domain circuit simulation, and electromagnetic field simulation using S-parameter. Therefore, the ADS software tool has found prominence in the design of RF electronic products in various commercial applications such as electric vehicle, mobile phones, wireless networks, satellites communication, and radar systems.

The tool has also been used for power converter design. The software provides various features such as S-parameter simulation and built-in mathematical functions that can help the designer in performing complex analysis from simulation data. The tool also has the feature of tuning, wherein the component parameters can be changed and results viewed in real time without having to rerun the simulation. This allows designers minimum design turnaround times.

Figure 6.9 shows the EGB structure with two layers. One of the layers is a solid plane layer, and the other is a patterned layer. The overall dimension of the whole structure is 90 mm × 90 mm with nine unit cells. Each unit cell is of 30 mm × 30 mm in size. The air gap of the center area of the unit cell is 0.4 mm and 0.2 mm projection with length of 5 mm and in the peripheral 0.1 mm gap with 6 mm projection. The whole unit cell is symmetric on all sides.

The board material FR-4 is used with its dielectric constant as 4.6 and its layer thickness as 2mm. The values are entered in the software, and the cross section of the printed circuit board is set up in the stack-up editor for two layers, as shown in Figure 6.10.

The unit cell is step-and-repeated as 3 × 3 matrix, as shown in Figure 6.11, and two ports are enabled to provide the excitation signal.

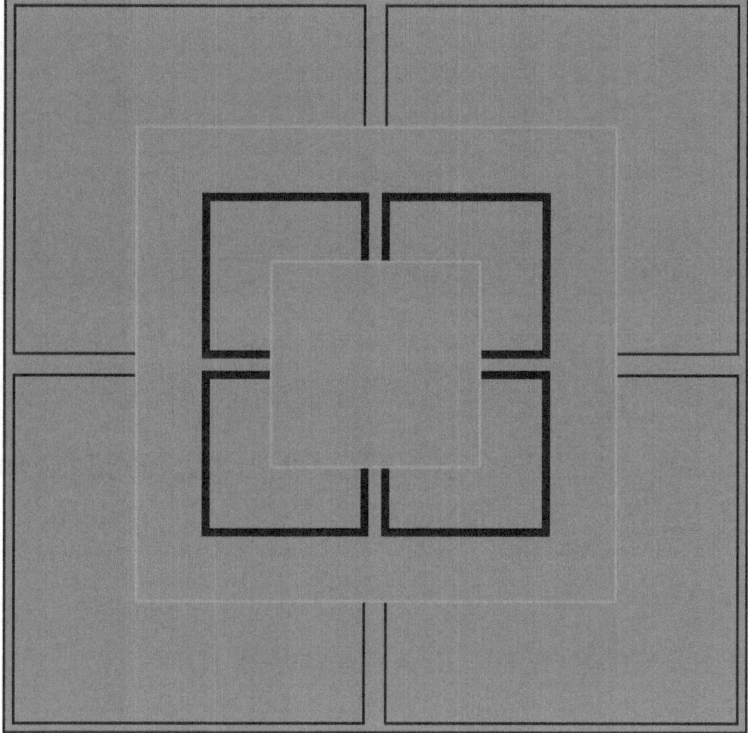

FIGURE 6.9   Unit structure of EGB.

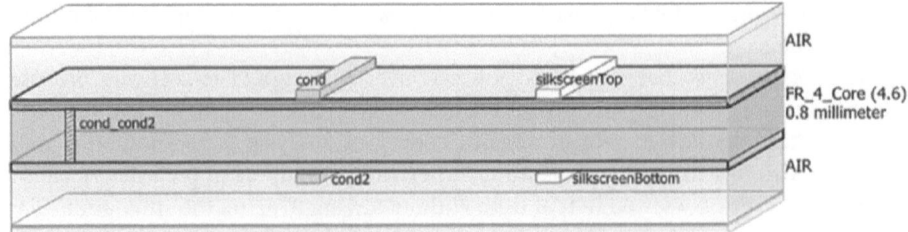

FIGURE 6.10   Substrate setup in ADS.

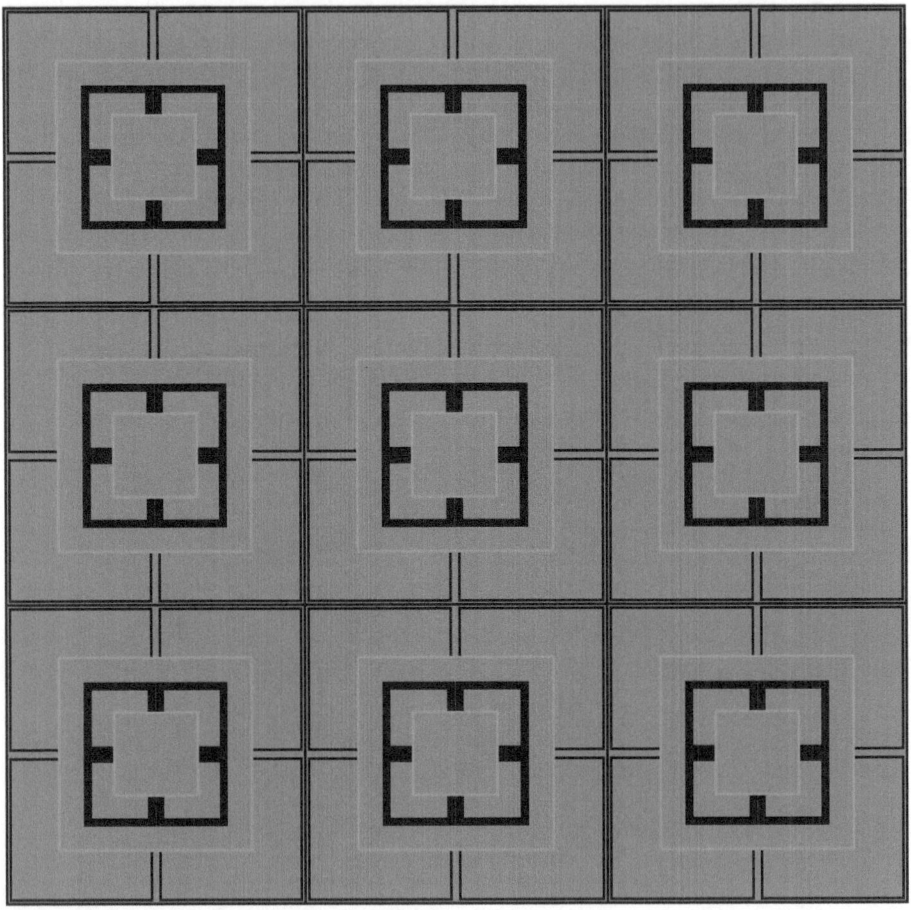

FIGURE 6.11   EBG pattern.

6.12 SIMULATION OF EBG IN A SOFTWARE TOOL

In the simulation tool, it is possible to select one of the EM simulators. They are momentum microwave, momentum RF, and FEM. Before running the simulation, the EM setup has to be done for the substrate, ports, and frequency plan. The

S-parameter model can be extracted by an EM simulator for the layout. The substrate setting constitutes the layer count of the PCB and the composition of each layer with the material parameters of the layers. The software tool enables the modeling of arbitrary 3D shapes with full-wave 3D electromagnetic field solver and doing completely automated meshing process and possess convergence capabilities. The tool automatically divides the geometric model into a large number of tetrahedrals, wherein four equilateral triangles are made by a single tetrahedron. A mesh is a basis from which a simulation begins. The port can be set for providing the excitation to the signal.

In this EBG design, two ports, P1 and P2, are set in the port editor. The port editor in the software tool enables the adding of the associated port details such as reference ground layer and feed type. Ground layer information is necessary for the ports having a negative pin. In this case, the reference layer has to be selected for a minus pin. Once placed in the layout, the port name cannot be edited in the port editor. Also, the tool allows the user to enter the feed type such as TML, SMD, DeltaGap. Generally, Auto can be selected, and the simulator will automatically select the required parameter. Ports are defined in the tool by adding pins in the layout with the required location.

Alternatively, this can be imported from other matching designs in order to avoid redundant work. Another important process in defining the port is to specify the usage of ports in the editor window of the EM simulation. Usually, the ports are added in the midpoint of the object surface. These definitions are saved in the layout.

In case of changes in layout, the tool indicates that changes in the ports' definitions and designs can be refreshed to modify the affected ports. Once the port definitions are done in the port editor, the EM setup window is reflected for the port definition. However, it is advised to save the settings before doing the simulations. Positive and negative terminal connections are established by these definitions. The following have to be specified:

- Port name
- Layer name where the port is to be inserted
- Net name
- Connected terminal name
- Purpose of the port
- X and Y locations in the layout
- Number specification of the port
- Pin-type specification

The natural field patterns are calculated by the 2D solver, which is present inside a transmission line structure for the same cross section. Here, the port and the resulting 2D field patterns serve as boundary conditions to the full-wave 3D problem. It is assumed that every port is inserted and linked with a uniform waveguide, and it is assumed to lie on the z-plane. The field associated with traveling waves propagating along the waveguide to which the port is connected is known as the excitation field. The field pattern can be calculated using Maxwell's equations. Here it is considered to be two layers and to have a frequency range from 0 to 25 GHz with the help of the EM simulator.

The EM set up can be done in the tool using the EM setup window. Here it is possible to feed the simulator with the following information:

- Setup type for simulation or cosimulation
- Type of simulator momentum or FEM
- Layout information
- Details of circuit partitioning
- Substrate view
- Port information view (specified in the port editor)
- Frequency plan definition
- Simulation option definition
- Definition of output plan
- Simulation resource specification
- EM model specification details

The definition of the output plan includes the following parameters:

- Physical model specification setting for momentum
- Physical model specification setting for FEM
- Setting for preprocessor
- Mesh setting for momentum
- Mesh setting for FEM
- Solver setting
- Expert option definition setting

The detailed information for every port is inserted in the layout for measuring the S-parameter. For editing the port, the port editor needs to be opened.

The definition of the required frequency plan can be set in the EM setup. Here it is possible to feed the multiple frequency plans for doing the simulation. Also, for every plan set, it is possible to define the solution to be found either for a range of frequencies or for a specified point of frequency. However, the simulation is executed as single run.

In order to feed the new frequency plan, the following detail has to be entered:

- Type—gives the frequency type:
  - Adaptive
  - Linear
  - Log
  - Single
  - Dec
  - SMPS
- Fstart—gives the start frequency in Hz, KHz, MHz, GHz, THz
- Fstop—gives the stop frequency in Hz, KHz, MHz, GHz, THz
- Npts—Number of frequencies to be specifies here for simulation. In the case of the adaptive frequency type, the maximum number of samples is taken.
- Step—gives the step value in Hz, KHz, MHz, GHz, THz
- Enabled—For enabling the frequency plan

In case of the SMPS type of frequency plan, it is required to enter the values for switching frequency, rise time, fall time, and ripple frequency. However, it is possible to delete the frequency plan in the setup window itself by selecting the necessary item.

Two ports are set using the port editor for exciting the signal. After the setup of all the required parameters such as frequency range and stack-up details, the simulation can be performed. The results are plotted in the form of graphs using the S-parameter. The Figure 6.12 shows the S-parameter simulation measurement output of the designed EBG structure, where it is observed that there is around a 60 dB reduction from the S21 measurement.

Figure 6.13a shows the value of S11 (dB), which is the input reflection gain. Figure 6.13d shows S22 (dB), which is the output reflection gain, and Figure 6.13b and c shows S21 (dB), forward gain, and S12 (dB), reverse gain, respectively. S21 and S12 represent forward and return gains (or losses) when the port network is terminated in reference impedance, and the input and output reflection gains are S11 and S22, respectively, which are negative decibel numbers.

In order to compare the effect of the application of the EBG structure, a continuous plane is constructed. Ports are set in the same location as in the EBG structure as shown in Figure 6.14. The parameter of the EM simulation is set as it is done for the EBG structure, the simulation is run in the ADS simulator, and the results are plotted. The S-parameter is analyzed for electromagnetic noise, and the results are shown in Figure 6.15.

Figure 6.15a shows the value of S11 (dB), which is input reflection gain, Figure 6.15d shows S22 (dB), which is output reflection gain, and Figure 6.15b and c show

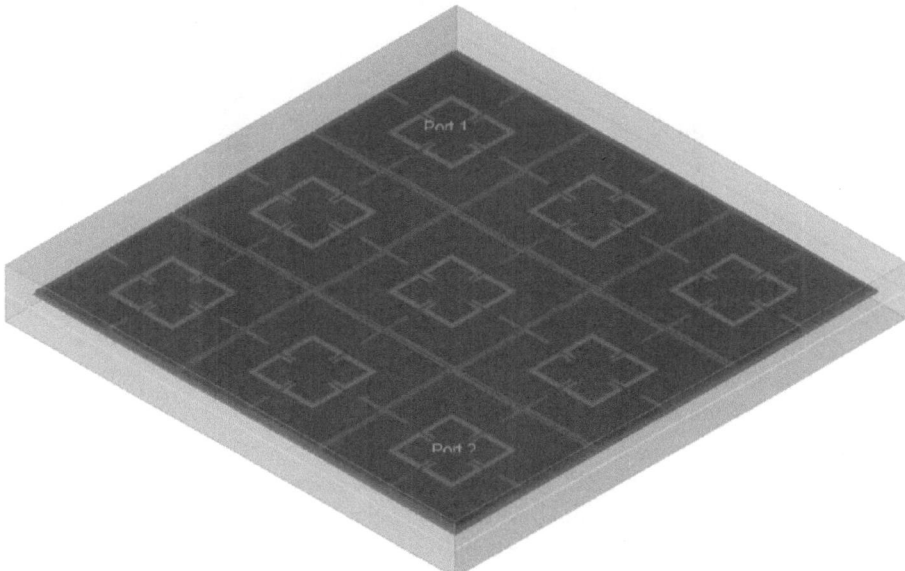

FIGURE 6.12   3D view of EBG simulation with ports.

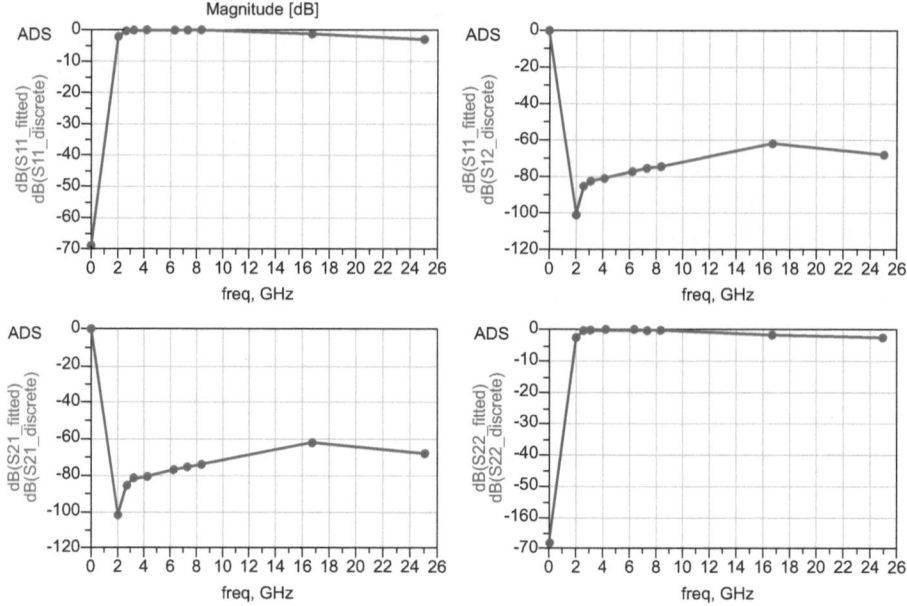

FIGURE 6.13 S-parameter output.

FIGURE 6.14 Continuous plane with ports excited.

S21 (dB), forward gain, and S12 (dB), reverse gain, respectively. It is observed that the noise level of −20 dB and the difference between both simulations are found and that around 40 dB, noise reduction is achieved with the application of the band gap structure. The depth in dB can be enhanced by altering the gap in the structure.

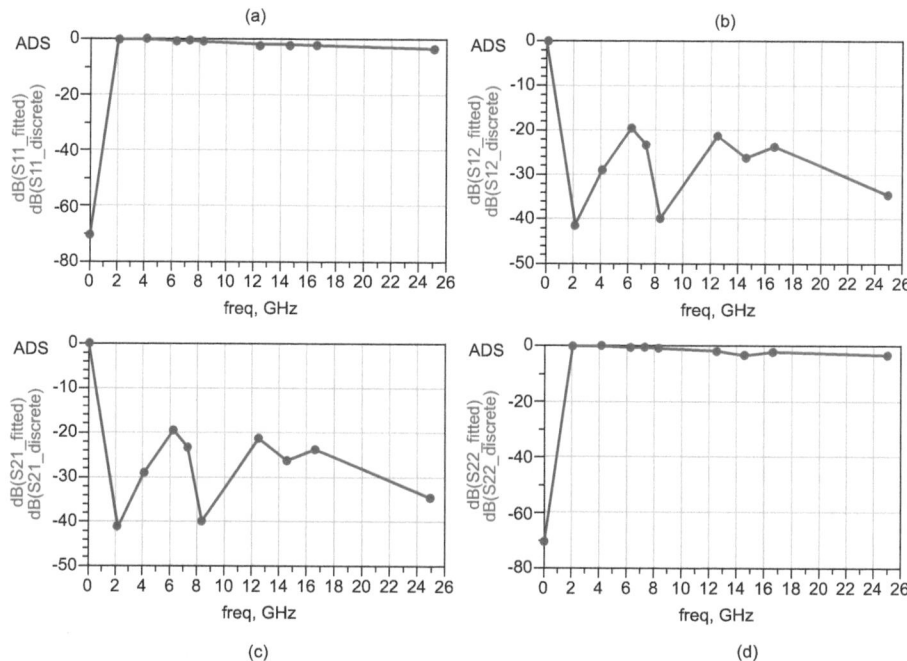

FIGURE 6.15   Continuous plane simulation S-parameter results.

The EBG structure was designed in order to mitigate the noise, and the same is simulated to realize the band gap noise ranging in the frequency from 0 to 25GHz with a depth of suppression as at −60 dB. It is compared with the continuous plane simulation results with the same parameter using the ADS simulator tool. From the analysis of the result, EBG structure acts as a filter that is used to mitigate noise levels.

The depth in dB level can be enhanced by altering the air gap in the unit cell structure. Also, the signal integrity with impedance discontinuity performance can also be done in the future for high-speed digital devices. The EBG structure can be manufactured, and the EMI measurement results using analyzers can be compared with the simulation results.

REFERENCES

[1] Bruce Archambeault, and Sam Connor, "Review of printed-circuit-board level EMI/EMC issues and tools", IEEE Trans. Electromagn. Compat., vol. 52, no. 2, May 2010.

[2] Shih-Vi Yuan, We-Bing SU, and Hao-Ping Ho, "A software technique for EMI optimization", 978-1-4577-1559-4/12/$26.00 ©2012 IEEE, https://ieeexplore.ieee.org/document/6237875

[3] Mandeep Kaur, Shikha Kakar, and Danvir Mandal, "Electromagnetic interference", 978-1-4244-8679-3/11/$26.00 ©2011 IEEE, https://ieeexplore.ieee.org/document/5941844

[4] Chuic Song et al., "Modeling of conducted EMI noise in an Automotive LED driver module with DC/DC converters", IEEE Int. Symp. Electromagn. Compat., 2019, https://ieeexplore.ieee.org/abstract/document/8872093

[5] A. Tsukioka et al., "Simulation techniques for EMC compliant design of automotive IC chips and modules," 2017 Int. Symp. Electromagn. Comp.—EMC EUROPE, Angers, pp. 1–6, 2017.

[6] Richard Lee Ozenbaugh, "EMI filter design", Marcel Dekker Inc., 2001.

[7] www.keysight.com

[8] CISPR25Standard, https://webstore.ansi.org/Standards/IEC

[9] E. Bogatin, "Signal and power integrity-simplified," in Plastics, 2nd ed. New Jersey: Prentice-Hall, 2009, pp. 188–189

[10] S. Van Den Berghe, F. Olyslager, D. De Zutter, J. De Moerloose, and W. Temmerman, "Study of the ground bounce caused by power plane resonances," IEEE Trans. Electromagn. Compat., vol. 40, no. 2, pp. 111–119, May 1998.

[11] T. L. Wu, S. T. Chen, J. N. Huang, and Y. H. Lin, "Numerical and experimental investigation of radiation caused by the switching noise on the partitioned DC reference planes of high-speed digital PCB," IEEE Trans. Electromagn. Compat., vol. 46, no. 1, pp. 33–45, Feb. 2004.

[12] T. Kamgaing, and O. M. Ramahi, "A novel power plane with integrated simultaneous switching noise mitigation capability using high impedance surface," IEEE Micro. Wireless Compon. Lett., vol. 13, pp. 21–23, Jan. 2003.

[13] T. L. Wu, Y. Y. Lin, C. C. Wang, and S. T. Chen, "Electromagnetic bandgap power/ground planes for wideband suppression of ground bounce noise and radiated emission in high-speed circuits," IEEE Trans. Microw. Theory Tech, vol. 53, no. 9, pp. 2935–2942, Sep. 2005.

[14] T. L. Wu, C. C. Wang, Y. H. Lin, T. K. Wang, and G. Chang, "A novel power plane with super-wideband elimination of ground bounce noise on high speed circuits," IEEE Micro. Wireless Compon. Lett., vol. 15, no. 3, pp. 174–176, Mar. 2005.

[15] X. H. Wang, B. Z. Wang, Y. H. Bi, and W. Shao, "A novel uniplanar compact photonic bandgap power plane with ultra-broadband suppression of ground bounce noise," IEEE Micro. Wireless Compon. Lett., vol. 16, no. 5, pp. 267–268, May 2006.

[16] J. Qin, and O. M. Ramahi, "Ultra-wideband mitigation of simultaneous switching noise using novel planar electromagnetic bandgap structures," IEEE Micro. Wireless Compon. Lett., vol. 16, no. 9, pp. 487–489, Sept. 2006.

[17] M. S. Zhang, Y. S. Li, C. Jia, L. P. Li, and J. Pan, "A double-surface electromagnetic bandgap structure with one surface embedded in power plane for ultra-wideband SSN suppression," IEEE Micro. Wireless Compon. Lett., vol. 17, no. 10, pp. 706–708, Oct. 2007.

[18] S. H. Joo, D. Y. Kim, and H. Y. Lee, "A S-bridged inductive electromagnetic bandgap power plane for suppression of ground bounce noise," IEEE Micro. Wireless Compon. Lett., vol. 17, no. 10, pp. 709–711, Oct. 2007.

[19] L. Li, Q. Chen, and K. Sawaya, "Ultra-wideband suppression of ground bounce noise using novel uniplanar compact electromagnetic bandgap structure," in Proc. 2008 IEEE AP-S USNC/URSI Symp., San Diego, CA, Jul. 5–12, 200.

[20] T. E. Moran, K. L. Virga, G. Aguirre, and J. L. Prince, "Methods to reduce radiation from split ground planes in RF and mixed signal packaging structures," IEEE Trans. Adv. Pack., vol. 25, no. 3, pp. 409–416, Aug. 2002.

[21] J. Qin, O.M. Ramahi, and V. Granatstein, "Novel planar electromagnetic band gap structures for mitigation of switching noise and EMI reduction in high-speed circuits," Trans. Trans. Electromagnet. Compat., vol. 49, no. 3, pp. 661–669, 2007.

[22] J. Li, J. Mao, S. Ren, and H. Zhu, "Embedded planar EBG and shorting via arrays for SSN suppression in multilayer PCBs," IEEE Antennas Wireless Propag. Lett., vol. 11, pp. 1430–1433, Nov. 2012.

[23] C. Shen, S. Chen, and T. Wu, "Compact cascaded-spiral-patch EBG structure for broadband SSN mitigation in WLAN applications," IEEE Trans. Microw. Theory Techn., vol. 64, no. 9, pp. 2740–2748, Sep. 2016.

[24] T. Wang, C. Hsieh, H. Chuang, and T. Wu, "Design and modeling of a stopband-enhanced EBG structure using ground surface perturbation lattice for power/ground noise suppression," IEEE Trans. Microw. Theory Techn., vol. 57, no. 8, pp. 2047–2054, Aug. 2009.

[25] H.-R. Zhu, and J. F. Mao, "Localized planar EBG structure of CSRR for ultrawide-band SSN mitigation and signal integrity improvement in mixed signal systems," IEEE Trans. Compon. Packag. Manuf. Technol., vol. 3, no. 12, pp. 2092–2100, Dec. 2013.

[26] Mahmoud, S. F., "A new miniaturized annular ring patch resonator partially loaded by a meta material ring with negative permeability and permittivity," IEEE Antennas Wirel. Propag. Lett., vol. 3, no. 1, pp. 19–22, 2004.

[27] M.-J. Gao, L.-S. Wu, and J. F. Mao, "Compact notched ultra-wideband bandpass filter with improved out-of-band performance using quasi electromagnetic band gap structure," Progress In Electromagnetics Research, vol. 125, pp. 137–150, 2012.

[28] Kim, M., and D. G. Kam, "Wideband and compact EBG structure with balanced slots," IEEE Trans. Compon. Packaging Manuf. Technol., Vol. 5, No. 6, 818–827, Jun.2015.

[29] Gonzalo, R., P. de Maagt, and M. Sorolla, "Enhanced patch-antenna performance by suppressing surface waves using photonic-bandgap substrates," IEEE Trans. Microw. Theory Tech., vol. 47, no. 11, 2131–2138, Nov. 1999.

[30] Yang, F. and Y. Rahmat-Samii, Electromagnetic Band Gap Structures in Antenna Engineering, Cambridge University Press, 2009, https://www.cambridge.org/core/books/electromagnetic-band-gap-structures-in-antenna-engineering/7F6D5B895E783 25CBFB22B5D1E970A4D

[31] Yang, F. and Y. Rahmat-Samii, "Mutual coupling reduction of microstrip antennas using electromagnetic band-gap structure," Proc. IEEE AP-S Int. Symp. Dig., vol. 2, pp. 478–481, Jul. 2001.

[32] Yang, F., and Y. Rahmat-Samii, "Microstrip antennas integrated with electromagnetic band-gap (EBG) structures: A low mutual coupling design for array applications," IEEE Trans. Antennas Propag., vol. 51, no. 10, pp. 2939–2949, Oct. 2003.

[33] Alam, M. S., M. T. Islam, and N. Misran, "A novel compact split ring slotted electro-magnetic band gap structure for micro strip patch antenna performance enhancement," Prog. Electromagn. Res., vol. 130, pp. 389–409, 2012.

[34] Yang, F. and Y. Rahmat-Samii, "Reflection phase characterizations of the EBG ground plane for low profile wire antenna applications," IEEE Trans. Antennas Propag., vol. 51, no. 10, pp. 2691–2703, Oct. 2003.

[35] Abedin, M. F., M. Z. Azad, and M. Ali, "Wideband smaller unit-cell planar EBG structures and their application," IEEE Trans. Antennas Propag., vol. 56, no. 3, pp. 903–908, Mar. 2008.

[36] Xu, H.-J., Y.-H. Zhang, and Y. Fan, "Analysis of the connection section between K connector and microstrip with Electromagnetic Band gap (EBG) structure," Prog. Electromagn. Res., vol. 73, pp. 239–247, 2007.

[37] Sievenpiper, D., L. Zhang, R. F. J. Broas, N. G. Alexopolus, and E. Yablonovitch, "High-impedance electromagnetic surfaces with a forbidden frequency band," IEEE Trans. Microw. Theory Tech., vol. 47, no. 11, pp. 2059–2074, 1999.

7 Efficient Energy Conversion Techniques for PV-fed Induction Motor Drive in Irrigation Applications

Josephine Rathinadurai Louis
and Rachaputi Bhanu Prakash

CONTENTS

DOI: 10.1201/9781003203810-7

LEARNING OUTCOMES

At the end of this chapter, the reader will be able to understand:

* The need for motors and drives in irrigation applications.
* The role of converters used in motors and drives.
* The design of energy efficient converters for motors and drives.
* Optimization techniques for converter operation.

7.1 INTRODUCTION

Water supply to the farming lands is a primary requirement and is generally fed from a motor pump setup. Most Indian villages lacking the utility power supply to drive these motor-pump drives (MPDs). Also, many farmlands in India do not have electrical supply, where the utility grid is not viable to extend because of their remote locations. The solution to resolve this power deficit is using renewable sources as a source of electricity. As India is a rich source of solar energy, the vulnerability of PV installation is highly recommended. The MPD in the farming lands will be fed through a solar PV array (SPVA) through the converter. This will act as a primary power source to the MPD in remote locations or an auxiliary source of power with utility grid. The motive is to propose a low-cost PV-array-fed MPD through an energy efficient two-stage power electronic converter network. The selection of motor plays an important role in the case of effective operation, and hence the induction motor is chosen because of its rugged performance and longer life. The induction motor-pump drive (IMPD) drives the PV array through a two-stage conversion procedure. The first stage incorporates a DC–DC step-up converter to develop the adequate voltage at the DC side of the inverter. A basic DC–DC boost converter is employed for stepping up the voltage. For solar PV applications demanding variable duty cycle at constant frequency operation of a DC–DC boost converter, there are more chances for the DC–DC boost converter to enter an unstable region [1]-[3]. Hence the boost converter is further modified with a few changes to make it well suitable to higher duty cycle ratios [4]–[5]. Many other DC–DC converters, such as Luo converters, use multistage RLC combinations to increase the voltage gain to a much higher value without compromising in stability [6]–[7]. The voltage and current stress across the semiconductor devices in these Luo converters are very large and thus demand higher ratings of switches and diodes. On the other hand, SEPIC and Cuk converters are further modified to have lower voltage and current stress across the semiconductor devices [8]–[12]. This will increase the life span of the devices. The voltage gain of the SEPIC converter is increased by employing switched inductors. The modification of a SEPIC converter with switched inductors can bring about an increased voltage gain. These higher values of gain will not last at very high duty cycle ratios and enter unstable regions when this DC–DC converter is exposed to MPPT- (maximum power point tracker-) based duty cycle control. Hence these converters require further modifications to have a stable voltage gain at higher values of duty cycle. Using a voltage-lift switched inductor (VLSI) is an alternative to an inductor in order to obtain a large voltage ratio. The first stage of the two-stage converter is a novel modified SEPIC converter that employs a modified voltage-lift switched inductor (MVLSI) to generate large voltage values without losing stability at higher duty cycle ratios [13]. The converter is fed from an SPVA, and for the maximum power extraction, a Particle Swarm Optimization (PSO)-based MPPT method will be implemented to control the duty cycle of the converter [14]–[16]. The second stage of the conversion process requires a conventional three-phase VSI (voltage source inverter) to feed IMPD. As the vector control is superior in controlling induction motors, it is employed to control the voltage source inverter VSI [17]–[19]. But the same vector control algorithm fails to control the induction motor when it is driven from SPVA. The starting current demanded by the induction motor causes the PV array to operate in short circuit mode and thus fails to develop voltage across the PV array. As a result of low voltage across the induction motor, it fails to rotate [20]–[24]. Hence, the vector control requires modification that does DC voltage regulation also. DC voltage regulation is used as speed reference generation for vector control to govern the input side of the three-phase VSI [25]–[28]. The service of the DC bus voltage regulation itself is not sufficient for maximum utilization of SPVA. The reference to the DC bus voltage regulation is obtained from the PSO-MPPT algorithm to extract maximum power.

The chronology of the book chapter is as follows. The second section consists of the proposed model for irrigation applications employing an induction motor drive. This section consists of the design of SPVA, a modeling water pump equivalent, and selection of DC link capacitor. The third section explains the MPPT control algorithm that is employed for the two-stage conversion procedure. The fourth section explains the proposed novel nonisolated DC–DC converter, considering operation, derivation of voltage gain, state space modeling with parasitic elements, design of L and C elements, and comparison among the proposed novel topologies. The fifth section explains the vector control of the induction motor (VCIM) and its modification with SPVA as the source of electrical power. The sixth section elucidates the simulation outcomes of the projected converter of its viability as a SPVA-fed IMPD.

7.2 TWO-STAGE CONVERTER FOR PV-ARRAY-FED IMPD

The energy conversion includes a two-stage conversion procedure as shown in Figure 7.1.

The first stage includes a nonisolated modified SEPIC converter. The converter is chosen to generate a large voltage ratio without losing converter stability. Also, the converter includes a maximum power point tracking (MPPT) algorithm, where duty cycle is controlled to extract maximum power even after inclusion of partial shading effects from the PV array. Now, the output DC voltage obtained from modified SEPIC converter enters the stage 2 process. In stage 2, the output from the converter is given to a conventional three-phase VSI. Hence VSI employs a sensor-less vector control approach to drive the induction motor. It is known that, when induction motor is fed by PV array, a huge inrush current during the starting process is required. The lack of it leads to failure during starting, when fed from a PV source. Therefore, a proper DC bus regulation is necessary at the input of three-phase VSI to regulate the voltage across the PV array. As a result, smooth starting of the induction motor takes place. For a three-phase VSI to operate, a normal vector control is applied, as it is effective. But, since the PV array is used and DC regulation is incorporated, this normal vector control will no longer be efficient. Hence, to improve the performance and handle proper energy conversion, a novel approach in the vector control is attempted. The novel approach includes input from the MPPT controller from stage 1, and the input is maximum peak power voltage. For the vector control, a subcontroller generates the reference speed. Now this speed is compared with the actual speed making three-phase voltage source inverter to operate efficiently. As the induction motor drive includes maximum peak power voltage, the motor-pump combination can work effectively.

7.3 DC LINK VOLTAGE CALCULATION WITH SPVA DESIGN (SIMULATION DATA)

The power rating of the SPVA is chosen to control the induction motor as an alternative to the utility grid. The motor selected for the irrigation applications is 2.2 kW

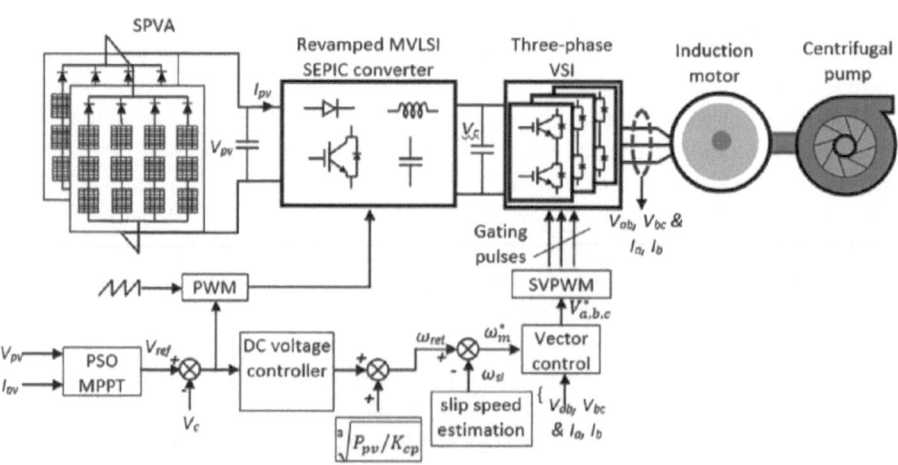

FIGURE 7.1 Two-stage converter for PV-array-fed IMPD.

(3 hp), 230 V, 1500 rpm induction motor. To drive the induction motor, the PV array should be able to develop 3 kW of power. The specifications of the solar panel are shown in the following table:

Parameter	Value (units)
Voltage at open circuit per module, V_{OC}	21.6 V
current at short circuit per module, I_{sc}	0.64 A
Maximum power point voltage per module, V_{mp}	17.6 V
Maximum power current per module, I_{mp}	0.58 A
Number of series connected modules, N_{se}	19
Number of parallel connected modules, N_{sh}	18

To meet the three-phase nominal voltage of 230 V, the voltage at the DC input side should be of $230*\sqrt{2} = 326$ V. So 19 series-connected panels are selected so that the voltage at maximum power of 17.6 V $*19 = 335$ V. The number of parallel-connected panels is 18 to have the current rating of the PV array = $0.58*18 = 10.4$ A. A power capacity of 326 V*10.4 A = 3390 W is chosen to drive the 2.2 kW induction motor drive.

The selection of the DC-link capacitor value is chosen by using fundamental component of frequency as follows [19], [20]:

$$\frac{1}{2}C_{dc}\left(V_{dc}^2 - V_{dca}^2\right) = 3aV_pI_t = 3 \times 1.2 \times 132.8 \times 10.4 \times 0.005 \tag{7.1}$$

Hence $C_{DC} = 2500$ μF.
Where:
V_{dca} = minimum permissible voltage at DC link capacitor during transient condition,
V_{dc} = DC link voltage, and
t = time required for a voltage to get reduced to minimum allowable DC link voltage.

DESIGN OF MOTOR PUMP (SIMULATION DATA)

1. The pump coupled to the motor has a relation between torque and speed developed in the motor [19], [20]: $T_L = K_{pump} \times \omega^2$. (7.2)
2. Where K_{pump} is the proportionality constant of the pump and is determined as $K_{pump} = P_{pump}/\omega^3$. P_{pump} is the mechanical input to the pump driven by the motor.
3. The same relationship is considered in the case of the induction motor feeding pump.

7.4 PARTICLE SWARM OPTIMIZATION ALGORITHM FOR MPPT OF SPVA

To feed power to govern the IMPD, the SPVA is required in the absence of a utility grid. The SPVA may be subjected to partial shading, and so the basic MPPT algorithms such as perturb and observe (P&O), incremental conductance (InC), hill climbing (HC), etc. may fail to function as a maximum power point (MPP) tracker. The particle swarm optimization (PSO) algorithm is one of the most popular algorithms to track MPP subjected to partial shading conditions as shown in Figure 7.2. The PSO algorithm is briefly explained next.

PSO is a simple nature-inspired method used for solving nonlinear analytical problems. PSO works on two principles: earning from the previous data, its primary principle, and imparting current data with the other swarm particles [29]. Agents in the PSO are called particles. All the particles follow the particles that exhibit the best performance, and all the agents should be directed toward the particle taking the optimum position. The particles tend to move until the stop criterion is met. The

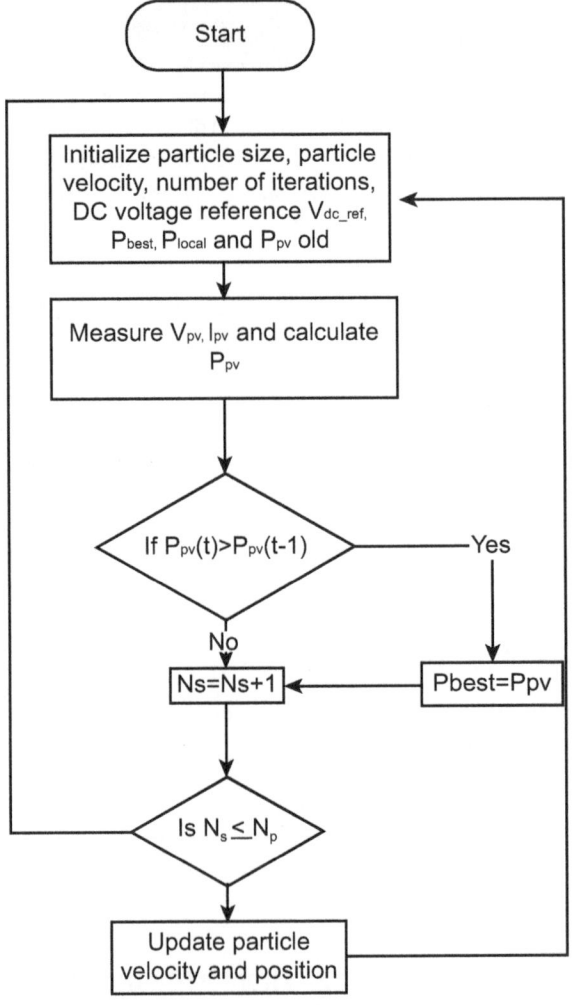

FIGURE 7.2 PSO algorithm used for MPPT under partial shaded conditions.

velocity and position update that tracks the MPPT with the objective function is as follows:

$$V_{k+1}^i = wV_k^i + C_1\left(P_{i\text{best}} - v_k^i\right) + C_2\left(g_{i\text{best}} - v_k^i\right) \quad (7.3)$$

And

$$v_{k+1}^i = v_k^i + V_{k+1}^i \quad (7.4)$$

where w is the inertia weight; C_1 and C_2 are arbitrary constants, the cognitive coefficient and the social coefficient, respectively.

The algorithm implementation involves the following chronology.

Step 1—Beginning of PSO: Select the particles in the exploration space with the chosen initial velocity in an arbitrary manner and quickly.

Step 2—Fitness assessment: The evaluation of fitness for individual particle can be done through the contribution of one agent elucidation to objective function.

Step 3—Updating global and local best particles: The generated fitness parameters are compared with the previous individual global and local best particle. The positions of the respective particles should also be updated accordingly.

Step 4—Updating of position and velocity: With Eqs. (3.1) and (3.2), updating of the positions and velocities of the particles is implemented.

Step5—Determining convergence: Once the criteria of the objective function are met, the procedure for the solution should be stopped. Otherwise, the process from step 2 should be repeated in the next iteration.

The PV array with the characteristics as shown in Figure 7.3 undergo different insolation at constant temperature. PSO can easily track the MPP by choosing random voltage points and finally settle at the MPP.

7.5 INCORPORATION OF MVLSI IN A REVAMPED SEPIC CONVERTER

7.5.1 Existing SEPIC Configurations and the Advancements in Their Topology

A modified SEPIC converter is realized from the SEPIC converter to minimize the voltage stress levels across the semiconductor devices, as shown in Figure 7.4.

The use of switched inductors is in practice to increase the voltage transfer function of the converter, and the voltage-lift switched inductor (VLSI) is a popular model as a replacement of an inductor. Figure 7.5 shows the various types of inductor models to be used in the place of inductors, out of which MVLSI is a modification to VLSI to attain high gain.

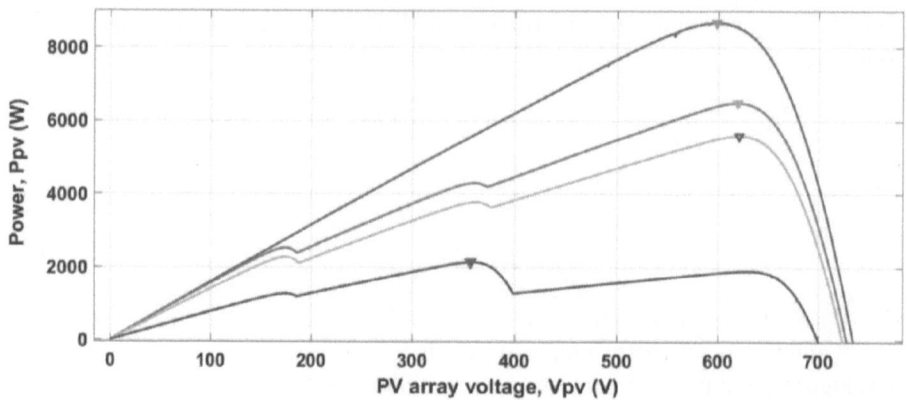

FIGURE 7.3 Power–voltage characteristics of SPVA when subjected to partial shading conditions.

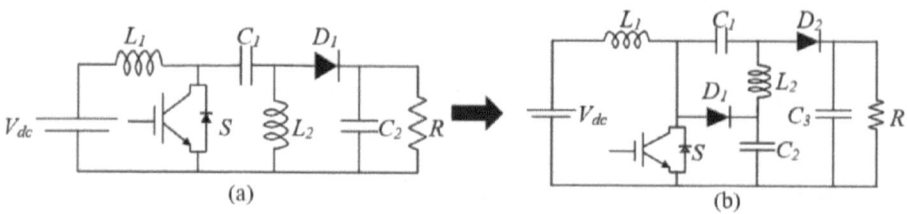

FIGURE 7.4 (a) SEPIC converter and (b) modified SEPIC converter.

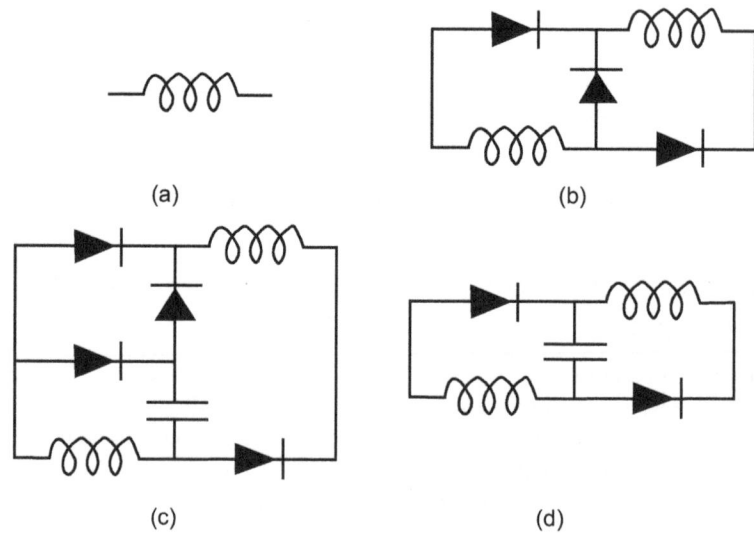

FIGURE 7.5 Types of inductor models: (a) inductor, (b) VLSI, (c) SI, (d) MVLSI.

The revamped SEPIC converter is incorporated with MVLSI to increase the stable gain and to reduce voltage stress across the switches. Three novel configurations are developed when this MVLSI is used as a replacement of the inductor in a revamped SEPIC converter and are depicted in Figure 7.6.

7.5.2 OPERATION OF THE CONVERTER

The main interpretation in the analysis of a revamped SEPIC converter is that the mode of operation is continuously conduction. It is meant to operate in two modes as it employs one switch, and one mode is with switch S in the "off" condition, while the other mode is with switch S in the "on" condition.

Mode I

Mode I operation is considered with the on-state of switch S. The proposed topologies during mode I operation are presented in Figure 7.5 for the three topologies I, II, and III. The charging of the inductors and capacitors in MVLSI takes place in this mode of operation through the diodes.

Mode II

Mode II operation is considered with the off condition of switch S. The proposed topologies during mode II operation are presented in Figure 7.5 for the

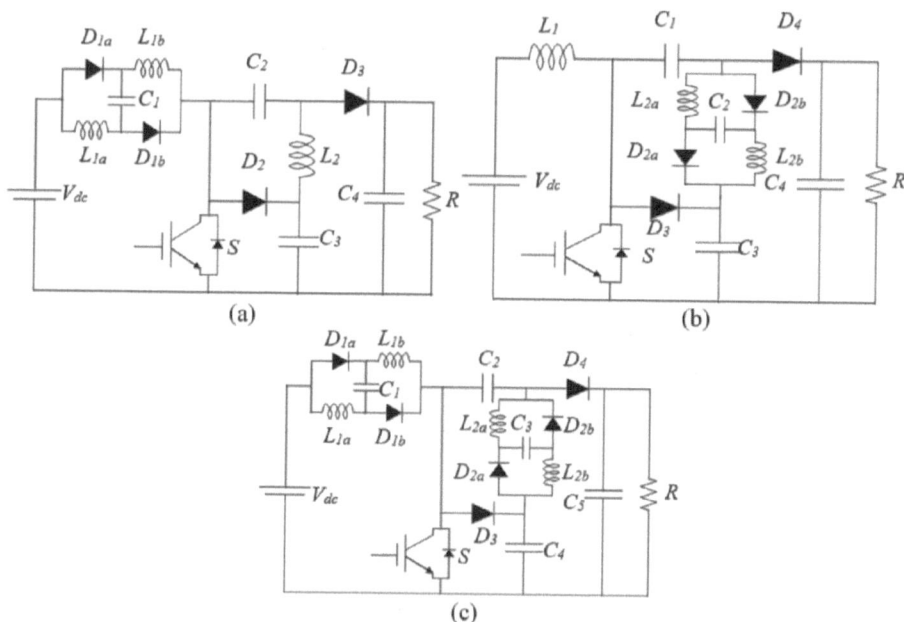

(a)

(b)

(c)

FIGURE 7.6 (a) Configuration 1, (b) configuration 2, and (c) configuration 3 of the MVLSI-based revamped SEPIC converter.

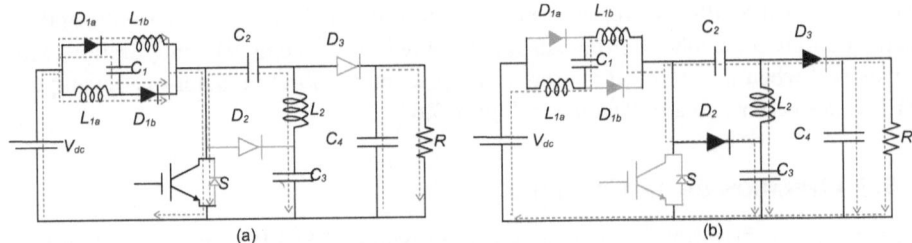

FIGURE 7.7   Operating modes of the projected topology.

three topologies I, II, and III. The discharging of the inductors and capacitors in MVLSI takes place in this state of process with the diodes in MVLSI in the off-state.

7.5.3 Derivation of Voltage Transfer Function of the Projected Configurations of MVLSI-Based Revamped SEPIC Converter

For Configuration 1

In the mode I of the configurations, the capacitor C_1 and the inductors L_{1a} and L_{1b} are charged in parallel with the DC input voltage source.

$$V_{L1a} = V_{L1b} = V_{C1} = V_{in} \tag{7.5}$$

$$L_{1a}\Delta i_{L1a} = L_{1b}\Delta i_{L1b} = V_{in}.DT \tag{7.6}$$

$$V_{C3} = V_{C2} + V_{L2}$$

replicates

$$V_{C3} = V_{C2} + \Delta i_{L2}/(DT) \tag{7.7}$$

During mode II, the inductors L_{1a}, L_{1b}, and the capacitor C_1 are discharged in series through the DC source. Thus

$$V_{C3} = V_{in} + V_{L1a} + V_{L1b} + V_{C1} \tag{7.8}$$

$$V_{C2} = V_{L2}$$

implies

$$V_{C2}.(1-D)T = L_2\Delta i_{L2} \tag{7.9}$$

From Eq. (7.5), Eq. (7.8) can be presented as

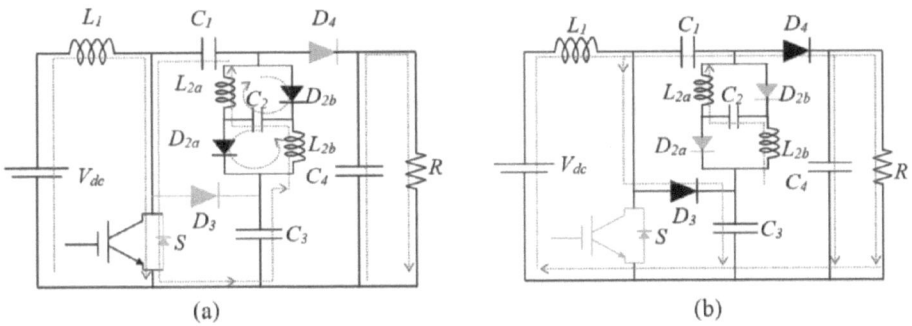

FIGURE 7.8   Operation of proposed configuration 2.

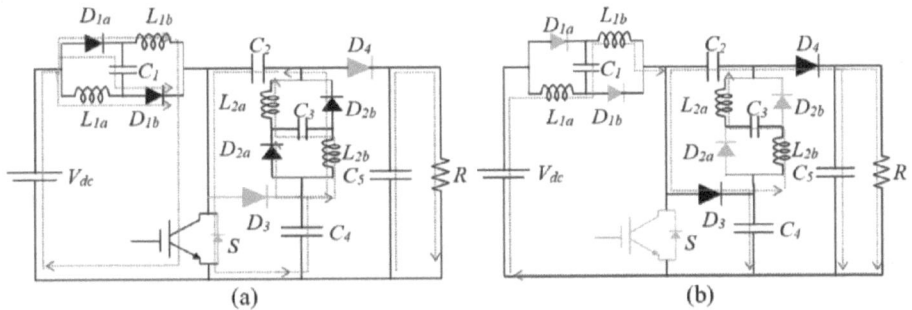

FIGURE 7.9   Operation of proposed configuration 3.

$$V_{C3} = 2V_{in} + 2V_{in}\left\{D/(1-D)\right\}$$

which replicates

$$V_{C3} = V_{in}\left(2/1-D\right) \tag{7.10}$$

By abridging Eq. (7.7) using (7.10),

$$V_{C3} = V_{C2}/D \tag{7.11}$$

$$V_o = V_{C2} + V_{C3}$$

which implies

$$V_o = (1+D)V_{C3} \tag{7.12}$$

From Eq. (7.10), Eq. (7.12) output voltage becomes

$$V_o = 2.V_{in}\left(1 + D/_{1-D}\right) \tag{7.13}$$

From Eq. (7.13), the voltage gain is obtained as

$$V_o/_{V_{in}} = 2.\left(1 + D/_{1-D}\right) \tag{7.14}$$

For Configuration 2 in mode I, the capacitor C_2 and inductors L_{2a} and L_{2b} are charged in parallel, sideways with C_3 by the capacitor C_1. Thus

$$V_{L2aon} = V_{L2bon} = V_{C2} \tag{7.15}$$

$$L_{2a}\Delta i_{L2a} = L_{2b}\Delta i_{L2b} = V_{C2}.DT \tag{7.16}$$

$$V_{L1on} = V_{in} \tag{7.17}$$

$$V_{C3} = V_{C2} + V_{C1}$$

replicates

$$V_{C3} = V_{C1} + L_{2a}\Delta i_{L2a}/_{DT} \tag{7.18}$$

During Mode I, the inductor L_1 and the capacitor C_3 are discharged through the input DC source. Thus

$$V_{C3} = V_{in} + V_{L1off}$$

$$V_{C3} = V_{in}\left(1/_{1-D}\right) \tag{7.19}$$

$$V_{C1} = V_{C2} + V_{L2aoff} + V_{L2boff} \tag{7.20}$$

$$V_{C1} = V_{C2}\left(1 + D/_{1-D}\right) \tag{7.21}$$

$$V_o = V_{C2} + V_{C4} \tag{7.22}$$

From Eqs. (7.19) and (7.21), voltage gain obtained for the configuration 2 by mathematical analysis can be written as

$$V_o/_{V_{in}} = 0.5\left(3 + D/_{1-D}\right) \tag{7.23}$$

During mode I in configuration 3, the capacitor C_1 and inductors L_{1a}, L_{1b} are charged in parallel connection with the input DC voltage source. Thus

$$V_{L1aon} = V_{L1bon} = V_{C1} = V_{in} \tag{7.24}$$

$$L_{1a}\Delta i_{L1a} = L_{1b}\Delta i_{L1b} = V_{in}.DT \tag{7.25}$$

$$\text{And } V_{C4} = V_{C2} + V_{L2a} \tag{7.26}$$

$$\text{But } V_{L2aon} = V_{C3}$$

During mode II, the capacitor C_1 and the inductors L_{1a} and L_{1b} are discharged in series with the DC source. Thus

$$V_{C4} = V_{in} + V_{C1} + V_{L1aoff} + V_{L1boff}$$

$$V_{C4} = 2V_{in}\left(\frac{1}{1-D}\right) \tag{7.27}$$

$$V_{C2} = V_{C3} + 2V_{L2aoff}$$

Substituting Eq. (7.27) in Eq. (7.26),

$$V_{C2} = V_{C3}\left(1 + \frac{D}{1-D}\right)$$

$$V_{C3} = V_{in} \tag{7.28}$$

$$V_o = V_{C2} + V_{C4}$$

$$V_o = V_{in}\left(\frac{3+D}{1-D}\right) \tag{7.29}$$

From Eq. (7.29), the voltage gain obtained for configuration 3 by the mathematical analysis is obtained as

$$V_o = V_{in}\left(\frac{3+D}{1-D}\right) \tag{7.30}$$

7.6 SELECTION OF CAPACITORS AND INDUCTORS IN THE DEVELOPED CONFIGURATIONS OF MODIFIED SEPIC CONVERTER

The design of the projected DC–DC converter configurations will be done by considering the following provisions:

Rated power of output load: $P_{out} = 2200$ W
Rated voltage across output load: $V_{out} = 350$ V
Rated SPVA voltage at standard conditions: $V_{pv} = 35$ V
Converter operating frequency: $f_{sw} = 25$ kHz

7.6.1 CHOICE OF DUTY CYCLE

The proposed three configurations of the modified SEPIC converter are tested to attain a minimum voltage gain of 10 V at an operating frequency of 25 kHz.

The required duty cycle for configuration 1 to have the voltage gain of 10 V is

$$D = \left.\left(V_0 - 2V_{in}\right)\middle/\left(2V_{in} + V_0\right)\right. = \frac{4}{6} = 66.67\% \tag{7.31}$$

The required duty cycle for configuration 2 to have the voltage gain of 10 V is

$$D = \left.\left(2V_0 - 3V_{in}\right)\middle/\left(2V_0 + V_{in}\right)\right. = \frac{17}{21} = 80.95\% \tag{7.32}$$

The required duty cycle for configuration 3 to have the voltage gain of 10 V is

$$D = \left.\left(V_0 - 3V_{in}\right)\middle/\left(V_0 + V_{in}\right)\right. = \frac{7}{11} = 63.63\% \tag{7.33}$$

7.6.2 Selection of Inductances

The average current flowing through the inductance L_1 and L_2 are equal, and so the inductances are determined as follows:

$$L_1\left(= L_{1a} = L_{1b}\right) = L_2\left(= L_{2a} = L_{2b}\right) = \left.V_{in}D\middle/f_{sw}\right. * \Delta i_L \tag{7.34}$$

The maximum allowed ripple in the current is chosen as it should not exceed 10% of the input current. For the load of 2200 W, the input current at 35 V will be approximately 65 A. To have a voltage gain of 10 V, the minimum duty cycle is selected and is of 63.63% obtained for configuration 3, which is used to find the minimum inductance.

$$L_1 = L_2 = \frac{V_{in}D_{min}}{f_{sw} * \Delta i_L} \geq 137\,\mu H \tag{7.35}$$

7.6.3 Capacitance Design

Similarly, the maximum allowed ripple in the voltage across in all the capacitors will be 10% of the output voltage.

$$C = \frac{I_o}{f_{sw} * \Delta v_c} \geq 7.43\,\mu F \tag{7.36}$$

7.6.4 Calculation of Normalized Voltage Stress for the Three Topologies of Modified SEPIC Converter

The voltage stress across diodes and switches in the DC–DC converter during their off-state are significant in their design as the reliability of the DC–DC converter depends on them.

7.6.5 CALCULATION OF NORMALIZED VOLTAGE STRESS OF CONFIGURATION 1

The normalized voltage stress across diode and switch devices of configuration 1 are determined as given. The voltage stress of the switch (V_{sw}) in its off-state of mode II is the same as the voltage across the capacitor C_3. The voltage stress across the diode D_2 (V_{d2}) in its off-state of mode I is in parallel to capacitor C_3. Hence normalized voltage stress across the switch and diode D_2 is given by

[[COMP: Set subscripts sw, d2, in, L1a, c1, etc. in nonitalic]]

$$V_{sw} = V_{d2} = 2V_{in}\left(\frac{1}{1-D}\right) \tag{7.37}$$

The normalized stress level of voltage across the diodes D_{1a} and D_{1b} is as follows:

$$V_{d1a} = V_{c1} + V_{L1a}, V_{d1b} = V_{c1} + V_{L1b}$$

$$V_{d1a} = V_{d1b} = \left(\frac{V_{in}}{1-D}\right) \tag{7.38}$$

The normalized stress level of voltage across the diodes D_3 is as follows:

$$V_{d3} = V_o + V_{c2} - V_{c3}$$

$$V_{d3} = 2V_{in}\left(\frac{D}{1-D}\right) \tag{7.39}$$

7.6.6 CALCULATION OF NORMALIZED VOLTAGE STRESS OF CONFIGURATION 2

The normalized voltage stress across the semiconductor devices of configuration 2 is determined as given. The voltage stress of the switch (V_{sw}) in its off-state of mode II is the same as the voltage across the capacitor C_3. The voltage stress across the diode D_2 (V_{d2}) in its off-condition of mode-I is in parallel to capacitor C_3. Hence the normalized stress level of voltage across the switch and diode D_2 is given by

$$V_{sw} = V_{d2} = \left(\frac{V_{in}}{1-D}\right) \tag{7.40}$$

The normalized stress level of voltage across the diodes D_{2a} and D_{2b} is given by

$$V_{d2a} = V_{c2} + V_{L2a}, V_{d2b} = V_{c2} + V_{L2b}$$

$$V_{d2a} = V_{d2b} = \left(\frac{V_{in}}{2(1-D)}\right) \tag{7.41}$$

The normalized stress level of voltage across the diodes D_3 is given by

$$V_{d3} = V_o + V_{c2} - V_{c3}$$

$$V_{d3} = V_{in} \left({D}/{1-D} \right) \tag{7.42}$$

7.6.7 Normalized Voltage Stress Calculations of Configuration 3

The voltage stress across all diodes and switch of configuration 3 is determined as follows.

The voltage across the switch (V_{sw}) during its off-state in mode II is equal to the voltage across the capacitor C_4. The voltage across the diode D_3 (V_{d3}) during its off-state in mode I is in parallel to capacitor C_4.

Therefore, the normalized stress level of voltage across the diode D_2 and switch is given by

$$V_{sw} = V_{d3} = \left({2V_{in}}/{(1-D)} \right) \tag{7.43}$$

The normalized stress level of voltage across the diodes D_{1a} and D_{1b} is given by

$$V_{d1a} = V_{L1a} + V_{c1}, V_{d1b} = V_{L1b} + V_{c1}$$

$$V_{d1a} = V_{d1b} = \left({V_{in}}/{(1-D)} \right) \tag{7.44}$$

The normalized stress level of voltage across the diodes D_{2a} and D_{2b} is given by

$$V_{d2a} = V_{L2b} - V_{c3}, V_{d2b} = V_{L2b} - V_{c3}$$

$$V_{d2a} = V_{d2b} = \left({V_{in}}/{(1-D)} \right) \tag{7.45}$$

The normalized stress level of voltage across the diodes D_3 is given by

$$V_{d3} = V_o + V_{c2} - V_{c4}$$

$$V_{d3} = V_{in} \left({6+2D}/{1-D} \right) \tag{7.46}$$

7.6.8 Stability Level Determination of the Projected Topologies with Parasitic Elements

The stability limit of the converter in the practical scenario depends on the parasitic elements. The stability limit of the proposed topologies is determined by considering the parasitic elements of the ideal elements along with the following assumptions:

1. A loss-less switch is considered for the analysis.
2. All the inductors and capacitors have parasitic resistances in series with them.
3. The source resistance is neglected in order to concentrate on the converter alone.
4. The forward-biased condition of diode is indicated as the voltage drop in series with forward resistance.

The stability limit of the proposed topologies can be assessed using the state space method.

The representation of the state space method to assess the stability of the proposed converter is in the following form:

$$\dot{X} = AX + BU \text{ and } Y = CX + DU$$

Here, $A = A'd + A''(1-d)$, $B = B'd + B''(1-d)$, $C = C'd + C''(1-d)$

And $D = D'd + D''(1-d)$

where X is the state variable matrix, d is the operating duty ratio of the converter and is the ratio of on-time of the switch to total time period of the operating switch, and U is the input matrix.

A', B', C', D' are the state space matrices during the on-time of the switch, and A'', B'', C'', D'' are the state space matrices during the off-time of the switch. The corresponding circuit of configuration 1 for both modes of operation with parasitic elements are shown in Figure 7.10.

The matrices A', B', $Ç'$ D' of configuration 1 during mode I are given as follows:

$$A' = \begin{bmatrix} a'_{11} & a'_{12} & a'_{13} & 0 & 0 & 0 & 0 \\ a'_{21} & a'_{22} & a'_{23} & 0 & 0 & 0 & 0 \\ a'_{31} & a'_{32} & a'_{33} & 0 & 0 & 0 & 0 \\ 0 & 0 & 0 & a'_{44} & a'_{45} & a'_{46} & 0 \\ 0 & 0 & 0 & a'_{54} & 0 & 0 & 0 \\ 0 & 0 & 0 & a'_{64} & 0 & 0 & 0 \\ 0 & 0 & 0 & 0 & 0 & 0 & a'_{77} \end{bmatrix}, B' = \begin{bmatrix} \begin{pmatrix} b'_{11} & b'_{12} \\ b'_{21} & b'_{22} \end{pmatrix} & O_{2X2} \\ O_{5X2} & O_{5X2} \end{bmatrix}$$

$$C'T = \begin{bmatrix} O_{6X1} \\ C'_7 \end{bmatrix} \& D' = [0]_{1X4}$$

The matrices $O_{2X2}, O_{5X2} \, O_{6X1}$, and O_{1X4} are null matrices. The matrices A'', B'', C'', and D'' of configuration 1 during mode II are given as follows:

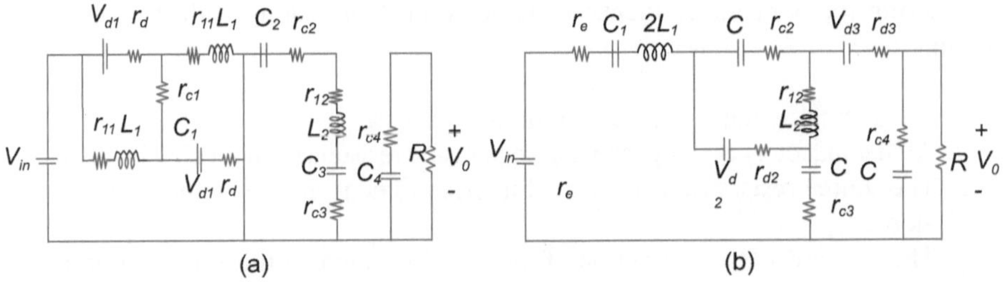

FIGURE 7.10 Two modes of operation of the proposed configuration 1 with parasitic elements.

$$
A'' = \begin{bmatrix}
a_{11} & 0 & a_{13} & a_{14} & a_{15} & a_{16} & a_{17} \\
a_{11} & 0 & a_{13} & a_{14} & a_{15} & a_{16} & a_{17} \\
a_{31} & 0 & 0 & 0 & 0 & 0 & 0 \\
a_{41} & 0 & 0 & a_{44} & a_{45} & a_{46} & a_{47} \\
a_{51} & 0 & 0 & a_{54} & a_{55} & a_{56} & a_{57} \\
a_{61} & 0 & 0 & a_{64} & a_{65} & a_{66} & a_{67} \\
a_{71} & 0 & 0 & a_{74} & a_{75} & a_{76} & a_{77}
\end{bmatrix}, \quad
B'' = \begin{bmatrix}
b_{11} & 0 & b_{13} & b_{14} \\
b_{11} & 0 & b_{13} & b_{14} \\
0 & 0 & 0 & 0 \\
0 & 0 & b_{43} & b_{44} \\
0 & 0 & b_{53} & b_{54} \\
0 & 0 & b_{63} & b_{64} \\
0 & 0 & b_{73} & b_{74}
\end{bmatrix},
$$

$$
\left(C'' \right)^{T} = \begin{bmatrix}
c_{11} \\
0 \\
0 \\
c_{14} \\
c_{15} \\
c_{16} \\
c_{17}
\end{bmatrix} \quad \text{and} \quad D''^{T} = \begin{bmatrix}
0 \\
0 \\
D_{13} \\
D_{14}
\end{bmatrix}
$$

The state variable matrix and the input matrix considered for configuration 1 are as follows:

$$X^T = \begin{bmatrix} i_{L1a} & i_{L1b} & V_{c1} & i_{L2} & V_{c2} & V_{c3} & V_{c4} \end{bmatrix}, U^T = \begin{bmatrix} V_{in} & V_{d1} & V_{d2} & V_{d3} \end{bmatrix}.$$

The arbitrary constants mentioned in these state matrices are presented in Appendix A1.

The equivalent circuit of configuration 2 with parasitic elements is shown in Figure 7.11.

The matrices A', B', C', and D' of configuration 2 during mode I are given as follows.

$$A'' = \begin{bmatrix} a'_{11} & 0 & 0 & 0 & 0 & 0 & 0 \\ 0 & a'_{22} & a'_{23} & a'_{24} & a'_{25} & a'_{26} & 0 \\ 0 & a'_{32} & a'_{33} & a'_{34} & a'_{35} & a'_{36} & 0 \\ 0 & a'_{42} & a'_{43} & a'_{44} & a'_{45} & a'_{46} & 0 \\ 0 & a'_{52} & a'_{53} & a'_{54} & a'_{55} & a'_{56} & 0 \\ 0 & a'_{62} & a'_{63} & a'_{64} & a'_{65} & a'_{66} & 0 \\ 0 & 0 & 0 & 0 & 0 & 0 & a'_{77} \end{bmatrix},$$

$$B'^T = \begin{bmatrix} b'_{11} & 0 & 0 & 0 & 0 & 0 & 0 \\ 0 & 0 & 0 & 0 & 0 & 0 & 0 \\ 0 & b'_{23} & b'_{33} & b'_{43} & b'_{53} & b'_{63} & 0 \\ 0 & 0 & 0 & 0 & 0 & 0 & 0 \end{bmatrix}, C'^T = \begin{bmatrix} O_{6X1} \\ C'_7 \end{bmatrix} \& D' = \begin{bmatrix} 0 \end{bmatrix}_{1X4}$$

The matrices $O_{2X2}, O_{5X2} O_{6X1}$, and O_{1X4} are null matrices.

The matrices A'', B'', C'', and D'' of configuration 1 during mode II are given as follows:

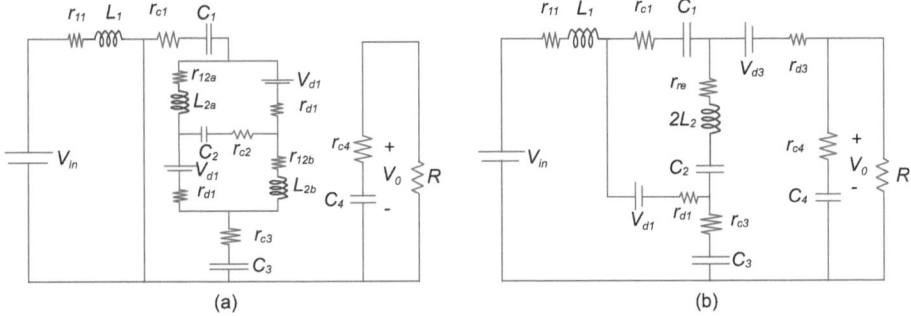

(a)   (b)

FIGURE 7.11 Two modes of operation of the proposed configuration 2 with parasitic elements.

$$
A'' = \begin{bmatrix}
a_{11} & a_{12} & 0 & a_{14} & 0 & a_{16} & a_{17} \\
a_{21} & a_{22} & 0 & a_{24} & a_{25} & a_{26} & a_{27} \\
a_{21} & a_{22} & 0 & a_{24} & a_{25} & a_{26} & a_{27} \\
a_{41} & a_{42} & 0 & a_{44} & 0 & a_{46} & a_{47} \\
0 & a_{52} & 0 & 0 & 0 & 0 & 0 \\
a_{61} & a_{62} & 0 & a_{64} & 0 & a_{66} & a_{67} \\
a_{71} & a_{72} & 0 & a_{74} & 0 & a_{76} & a_{77}
\end{bmatrix}, \quad
B'' = \begin{bmatrix}
b_{11} & b_{12} & 0 & b_{14} \\
0 & b_{22} & 0 & b_{24} \\
0 & b_{22} & 0 & b_{24} \\
0 & b_{42} & 0 & b_{44} \\
0 & 0 & 0 & 0 \\
0 & b_{62} & 0 & b_{64} \\
0 & b_{72} & 0 & b_{74}
\end{bmatrix},
$$

$$
(C'')^T = \begin{bmatrix}
c_{11} \\
c_{12} \\
0 \\
c_{14} \\
0 \\
c_{16} \\
c_{17}
\end{bmatrix}, \quad
D''T = \begin{bmatrix}
0 \\
D_{12} \\
0 \\
D_{14}
\end{bmatrix}
$$

The state variable matrix and the input matrix considered for configuration 2 are

$$
X^T = \begin{bmatrix} i_{L1} & i_{L2a} & i_{L2b} & V_{c1} & V_{c2} & V_{c3} & V_{c4} \end{bmatrix}, \quad U^T = \begin{bmatrix} V_{in} & V_{d1} & V_{d2} & V_{d3} \end{bmatrix}
$$

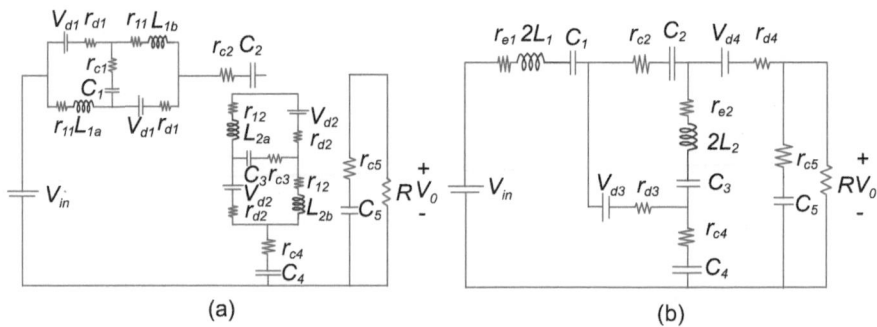

(a)   (b)

FIGURE 7.12 Two modes of operation of the proposed configuration 3 with parasitic elements.

The equivalent circuit of configuration 3 with parasitic elements is shown in the Figure 7.12.

The matrices A', B', C', and D' of configuration 3 during mode I are given as follows:

$$A' = \begin{bmatrix} a'_{11} & a'_{12} & a'_{13} & 0 & 0 & 0 & 0 & 0 & 0 \\ a'_{21} & a'_{22} & a'_{23} & 0 & 0 & 0 & 0 & 0 & 0 \\ a'_{31} & a'_{32} & a'_{33} & 0 & 0 & 0 & 0 & 0 & 0 \\ 0 & 0 & 0 & a'_{44} & a'_{45} & a'_{46} & a'_{47} & a'_{48} & 0 \\ 0 & 0 & 0 & a'_{54} & a'_{55} & a'_{56} & a'_{57} & a'_{58} & 0 \\ 0 & 0 & 0 & a'_{64} & a'_{65} & a'_{66} & a'_{67} & a'_{68} & 0 \\ 0 & 0 & 0 & a'_{74} & a'_{75} & a'_{76} & a'_{77} & a'_{78} & 0 \\ 0 & 0 & 0 & a'_{84} & a'_{85} & a'_{86} & a'_{87} & a'_{88} & 0 \\ 0 & 0 & 0 & 0 & 0 & 0 & 0 & 0 & a'_{99} \end{bmatrix},$$

$$(B')^T = \begin{bmatrix} b'_{11} & b'_{21} & b'_{31} & 0 & 0 & 0 & 0 & 0 & 0 \\ b'_{12} & b'_{22} & b'_{32} & 0 & 0 & 0 & 0 & 0 & 0 \\ 0 & 0 & 0 & b'_{43} & b'_{53} & b'_{63} & b'_{73} & b'_{83} & 0 \\ 0 & 0 & 0 & 0 & 0 & 0 & 0 & 0 & 0 \\ 0 & 0 & 0 & 0 & 0 & 0 & 0 & 0 & 0 \end{bmatrix}$$

$$C' = \begin{bmatrix} O_{1X8} & c'_9 \end{bmatrix} \& D' = \begin{bmatrix} O \end{bmatrix}_{1X5}$$

The matrices O_{1X8} and O_{1X5} are null matrices.

The matrices A'', B'', C'', and D'' of configuration 3 during mode II are given as follows:

$$A' = \begin{bmatrix} a_{11} & 0 & a_{13} & a_{14} & 0 & a_{16} & 0 & a_{18} & a_{19} \\ a_{21} & 0 & a_{13} & a_{14} & 0 & a_{16} & 0 & a_{18} & a_{19} \\ a_{31} & 0 & 0 & 0 & 0 & 0 & 0 & 0 & 0 \\ a_{41} & 0 & 0 & a_{44} & 0 & a_{46} & a_{47} & a_{48} & a_{49} \\ a_{41} & 0 & 0 & a_{44} & 0 & a_{46} & a_{47} & a_{48} & a_{49} \\ a_{61} & 0 & 0 & a_{64} & 0 & a_{66} & 0 & a_{68} & a_{69} \\ 0 & 0 & 0 & a_{74} & 0 & 0 & 0 & 0 & 0 \\ a_{81} & 0 & 0 & a_{84} & 0 & a_{86} & 0 & a_{88} & a_{89} \\ a_{91} & 0 & 0 & a_{94} & 0 & a_{96} & 0 & a_{98} & a_{99} \end{bmatrix},$$

$$(B'')^T = \begin{bmatrix} b_{11} & b_{11} & 0 & 0 & 0 & 0 & 0 & 0 & 0 \\ 0 & 0 & 0 & 0 & 0 & 0 & 0 & 0 & 0 \\ 0 & 0 & 0 & 0 & 0 & 0 & 0 & 0 & 0 \\ b_{14} & b_{14} & 0 & b_{44} & b_{44} & b_{64} & 0 & b_{84} & b_{94} \\ b_{15} & b_{15} & 0 & b_{45} & b_{45} & b_{65} & 0 & b_{85} & b_{95} \end{bmatrix}$$

$$C'' = \begin{bmatrix} C_1 & 0 & 0 & C_4 & 0 & C_6 & 0 & C_8 & C_9 \end{bmatrix} \text{ and } D'' = \begin{bmatrix} 0 & 0 & 0 & D_4 & D_5 \end{bmatrix}$$

The stable output for the maximum possible duty cycle is calculated by finding the eigenvalues of matrix A. The eigenvalues of state variable matrix A are nothing but the poles of the system. The eigenvalues are determined using the state variable matrix by varying the duty cycle of the converter. The practical values of the elements used in the proposed configurations are presented in Table 7.1. To develop a prototype and to test the process of the projected DC–DC converters, the input voltage is taken as 5 V for a required load voltage of 50 V as implemented in simulation. The obtained eigenvalues of each configuration of projected converter at the verge of stability is tabulated as shown in Table 7.2 by simulating them in MATLAB.

7.7 MODIFIED ENCODER LESS VECTOR CONTROL OF INDUCTION MOTOR DRIVE

7.7.1 ESTIMATION OF SPEED IN INDUCTION MOTOR DRIVE

Estimation of speed in the induction motor can be done without the help of a speed sensor or encoder, and this can be done with the help of measuring the instantaneous voltage and currents drawn by the induction motor.

TABLE 7.1

Revamped SEPIC Converter Specifications for the Proposed Configurations.

Symbol	Specification	Value(s)
V_{in}	SPVA voltage	5 V
V_{out}	DC–DC load voltage	50 V
P_o	Rated power	10 W
f_{sw}	Operating frequency of switch	25 kHz
L	Inductance	5 mH, 10 A
r_l	Inductor series resistance	13 mΩ
C	Capacitors	220 μF, 63V
r_c	Capacitor series resistance	11 mΩ
V_d	Diode voltage drop	0.7 V
r_d	Diode series resistance	35 mΩ
R	Load resistance	1 kΩ

TABLE 7.2
Eigenvalues of the Proposed Configurations at the Verge of Stability.

Revamped SEPIC converter configuration 1

Duty cycle	Eigenvalues (poles)	Duty cycle	Eigenvalues (poles)
0.895	−4.8199	**0.896**	−4.8311
	−0.6245		**0.6156**
	−0.012 + j0.232		−0.0119 + j0.0231
	−0.012 − j0.0232		−0.0119 − j0.0231
	−0.01 − j0.0288		**0.0099 +j 0.0227**
	−0.01 + j0.0288		**0.0099 − j0.0227**
	−0.0001		−0.0001

Revamped SEPIC converter configuration 2.

Duty cycle	Eigenvalues (poles)	Duty cycle	Eigenvalues (poles)
0.912	−1.2385	**0.913**	−1.2239
	−0.806 + j0.396		**0.797 + j0.391**
	−0.806 − j0.396		**0.797 − j0.391**
	−0.0663		**0.0708**
	−0.0001 + j 0.007		−0.0001 + j 0.007
	−0.0001 − j 0.007		−0.0001 − j 0.007
	0		0

Revamped SEPIC converter configuration 3

Duty cycle	Eigenvalues (poles)	Duty cycle	Eigenvalues (poles)
0.996	−8.8243	**0.997**	−8.8247
	−5.6094		−5.61
	−4.4065		−4.4087
	−0.000 + j0.135		**0.0001 + j0.135**
	−0.0001 − j0.135		**0.0001 − j0.135**
	−0.0004		−0.0004
	−0.0013		−0.0009
	−0.0010		−0.0010
	−0.0010		−0.0010

The transformation of captured three-phase voltages and currents are performed according to α-β frame of reference to determine the flux quantities from which the motor speed can be estimated.

$$v_\alpha = \frac{1}{3}\left(2v_a - v_b - v_c\right), \ v_\beta = \frac{1}{3}\left(v_b - v_c\right) \tag{7.47}$$

$$i_\alpha = \frac{1}{3}\left(2i_a - i_b - i_c\right), \ i_\beta = \frac{1}{3}\left(i_b - i_c\right) \tag{7.48}$$

where i_a, i_b, and i_c are balanced three-phase winding currents.

$$p(\Psi_\beta) = \left(v_\beta - R_s i_\beta \right), p(\Psi_\alpha) = \left(v_\alpha - R_s i_\alpha \right) \tag{7.49}$$

$$\Psi_s = \sqrt{\Psi_\alpha^2 + \Psi_\beta^2} \tag{7.50}$$

$$i_{qs} = i_\beta \left(\Psi_\alpha / \Psi_s \right) - i_\alpha \left(\Psi_\beta / \Psi_s \right) \tag{7.51}$$

$$i_{ds} = i_\beta \left(\Psi_\beta / \Psi_s \right) + i_\alpha \left(\Psi_\alpha / \Psi_s \right) \tag{7.52}$$

$$\omega_e = \left[\left(v_\beta - R_s i_\beta \right) \Psi_\alpha - \left(v_\alpha - R_s i_\alpha \right) \Psi_\beta \right] / \Psi_s^2 \tag{7.53}$$

where i_{qs} and i_{ds} are the components of stator currents in synchronized dq_0 reference frame with $\tilde{A} = 1 - L_m^2 / (L_s L_r), \tau_r = L_r / R_r, L_r, L_m,$ and L_{lr} are the rotor inductance, magnetizing inductance, and rotor leakage inductance, respectively. The parasitic resistances R_r and R_s refer the stator and rotor, respectively.

The rotational speed of the motor is determined as follows:

$$\omega_m = \omega_e - \omega_{sl} \tag{7.54}$$

The synchronous speed (ω_e) and slip speed (ω_{sl}) are predicted as follows:

$$\acute{E}_{sl} = \frac{\left(1 + \sigma S \tau_r \right) L_s i_{qs}}{\tau_r \left(\Psi_{ds} - \sigma L_s i_{ds} \right)} \tag{7.55}$$

7.7.2 FIELD-ORIENTED CONTROL (FOC) OF IMD

Figure 7.12 depicts the control scheme of the vector control method, used for controlling the flux and stator currents.

$$I_{dm}^* = I_{dm}^{e*} + \tau_r p I_{dm}^{e*} \tag{7.56}$$

in normal operation. The reference flux (Ψ_{ds}) is determined as

$$\Psi_{ds}^* = L_m I_{dm}^{e*} \tag{7.57}$$

An exciting current I_{ds}^* is considered output of the flux generated error is directed to the input of PI controller.

$$I_{ds(k)}^* = I_{ds(k-1)}^* + K_{p\Psi} \left\{ \Psi_{e(k)} - \Psi_{e(k-1)} \right\} + K_{i\Psi} \Psi_{e(k)} \tag{7.58}$$

where

$$\Psi_e = \Psi_{ds}^* + \Psi_{ds} \tag{7.59}$$

A decoupling component is introduced in the FOC for controlling the flux and torque individually by introducing the feedforwarded path. The decoupling component (I_{dcp}) of the current is explained here.

Reference Speed Generation

The PSO algorithm generates an output of V_{dc}^* to generate an error signal. This error signal is given to the input of a PI controller to generate a speed estimation ω_1. To meet the power requirement of the pump, this estimated speed ω_1 is added to the pump speed, which results in ω_{ref}.

7.8 MATLAB SIMULATION

7.8.1 Simulation of Predicting Output Voltage of Projected DC–DC Converter Topologies

The MVLSI-based revamped SEPIC DC–DC converter topologies are designed for 50 V output voltage and 5 V input voltage. The MOSFET is triggered with a duty cycle of 66.67% for configuration 1, 80.95% for configuration 2, and 25 kHz. The specifications and components used are listed in Table 7.1 for configurations 1, 2, and 3. The MVLSI-based revamped SEPIC converter is tested in the MATLAB/SIMULINK environment. Waveforms of load voltage and current waveforms of configuration 1, configuration 2 and configuration 3 are shown in Figure 7.14, Figure 7.15, and Figure 7.15, respectively. From the simulation graphs, the output voltage and current are constant around 50 V and 0.038 A with a ripple of 0.34 and 0.39% respectively, for all the configurations 1, 2, and 3.

The input voltage is taken to be 5 V, so the output voltage to be obtained is 50 V. A SIMULINK model is designed with the same specifications as those of the hardware circuitry. These simulation outcomes are compared with the hardware outcomes obtained from the hardware model.

(a)

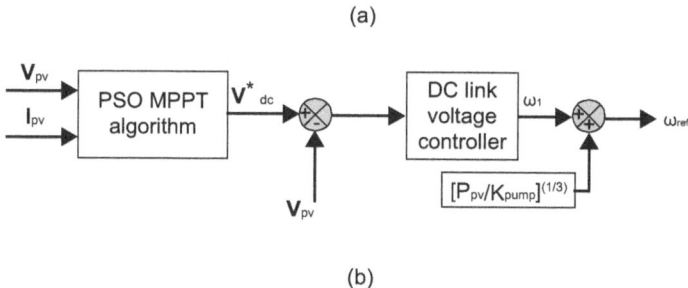

(b)

FIGURE 7.13 Generation of speed reference: (a) estimation of ω_1; (b) feedforward component of speed.

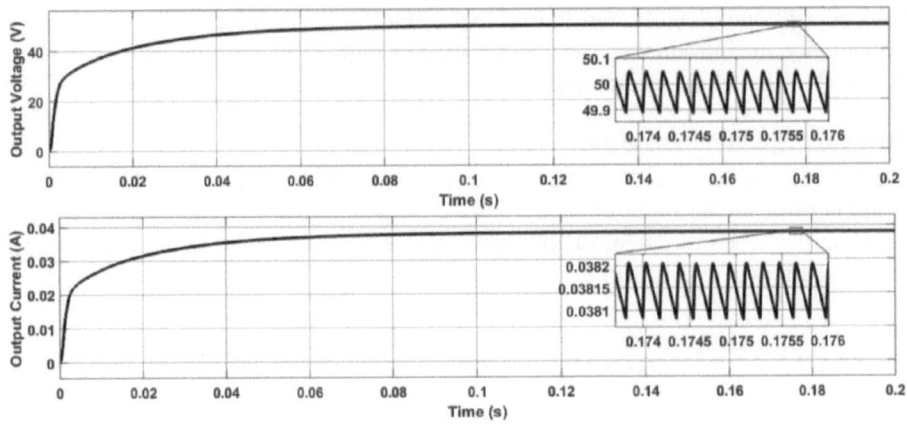

FIGURE 7.14   Waveforms of output voltage and current for configuration 1.

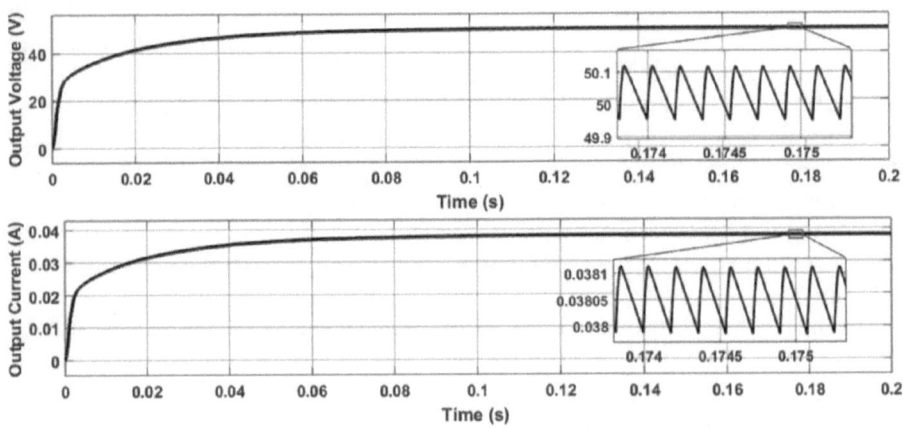

FIGURE 7.15   Waveforms of output voltage and current for configuration 2.

7.8.2 SIMULATION OF PV-ARRAY-FED IMPD THROUGH NOVEL TWO-STAGE CONVERTER

The major consideration is that the output power transfer to the load should be operated at MPP. The MPPT PSO method is implemented to transfer the peak power. The
simulations are performed for the proposed PV array configuration under the change
in insolation. The insolation changes are applied to the 34 series-connected panels
in such a way that 10 of the series panels are at the highest insolation. For example,
the first 10 modules are exposed to 1000 W/m² insolation, and 10 more modules
are exposed to 800 W/m², and the remaining 14 series-connected PV modules are
exposed to 700 W/m². The simulation is tested for the variation of insolation from
[1000 W/m², 800 W/m², 700 W/m²] of [7 6 6] series-connected panels to [700 W/

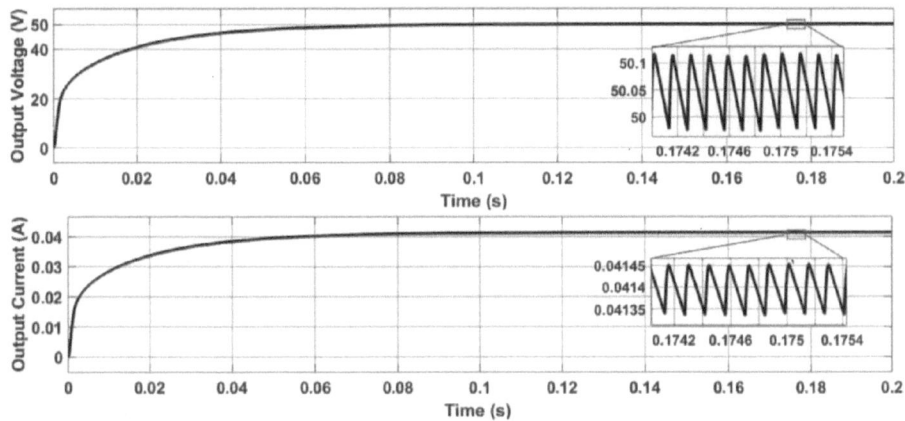

FIGURE 7.16 Waveforms of output voltage and current for configuration 3.

m², 500 W/m², 500 W/m²] of [7 6 6] series-connected panels. The maximum power varies with the insolation from 2.67 to 0.54 kW. All the simulation results are tested for the same decrease in insolation with constant temperature at 25°C. The change in insolation is applied to the converter, and the respective change in output power of DC–DC converter is obtained as shown in Figure 7.17. Only configuration 3 of the revamped SEPIC converter is considered for the analysis because of its stability at large duty cycle ratios. The regulation of DC voltage at the capacitor filter is maintained with the change in insolation as shown in Figure 7.18.

The power generated from the SPVA fed completely to the induction motor alone and the maximum power algorithm will be interfaced with vector control of the induction motor drive. The graphs shown in Figure 7.19 resemble the performance of the induction motor for the change in insolation. The variation of induction motor speed in reference to insolation of the PV array is significant, and so the impact is on the torque developed in the induction motor.

7.9 CONCLUSION

The book chapter introduced a novel two-stage conversion procedure to attain effective control of IMPD driven by a SPVA. The first stage involves a revamped SEPIC DC–DC converter topology employing MVLSI for the large output voltage requirement. The operation of the three configurations to derive the voltage gain was explained. The design of inductors and capacitors was done based on the chosen allowed ripple for the PV application. The maximum allowed duty cycle for the same converters was compared for the same values of inductors and capacitors. For the same gain of 10, configuration 3 was operated at a lower value of the duty cycle, and the stability limit of the converter was extremely large, which is almost in ideal state as the maximum allowed duty cycle obtained is 0.9996. This shows the feasibility of the developed DC–DC converter for PV array applications where the variable operating duty cycle is implemented. The second stage of the conversion involves

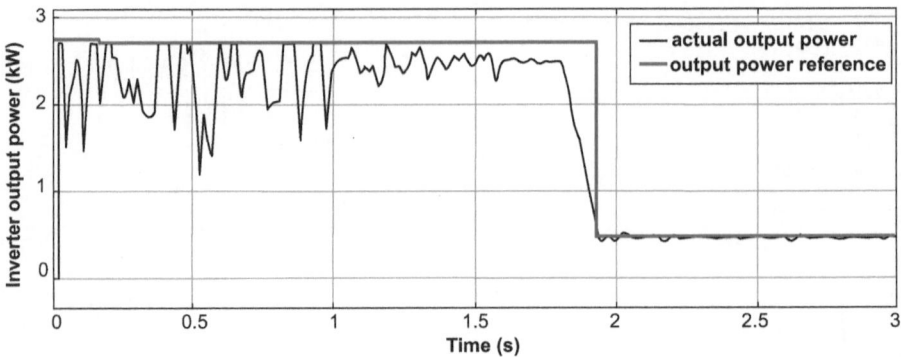

FIGURE 7.17   PSO-MPPT algorithm implementation to deliver the output power for change in insolation at 1.8 seconds.

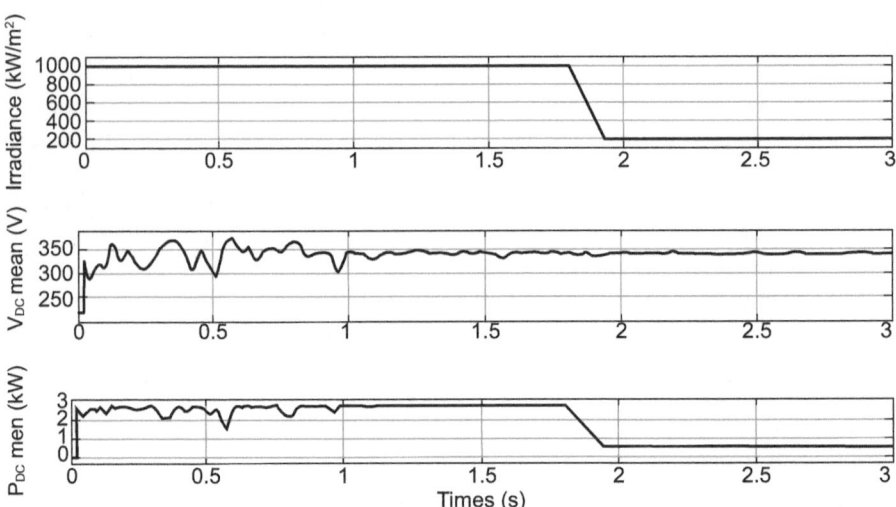

FIGURE 7.18   Variation of output power after the boost converter with respect to decrease in insolation at 1.8 seconds.

a three-phase inverter, controlled through PWM, generated from modified vector control of the IMD. The modification in vector control of the IMD was implemented by incorporating DC voltage regulation and a MPPT algorithm. A PSO-based MPPT algorithm was chosen for the operation of the two-stage converter. The simulation was performed on the two-stage converter for the decrease in insolation, and a proportional change in the performance of the induction motor is attained. It has been maintained that the output power generated is equal to the peak power point of SPVA at the respective insolation levels. Thus a new two-stage converter is proposed for an effective energy conversion to meet the requirement of irrigation applications.

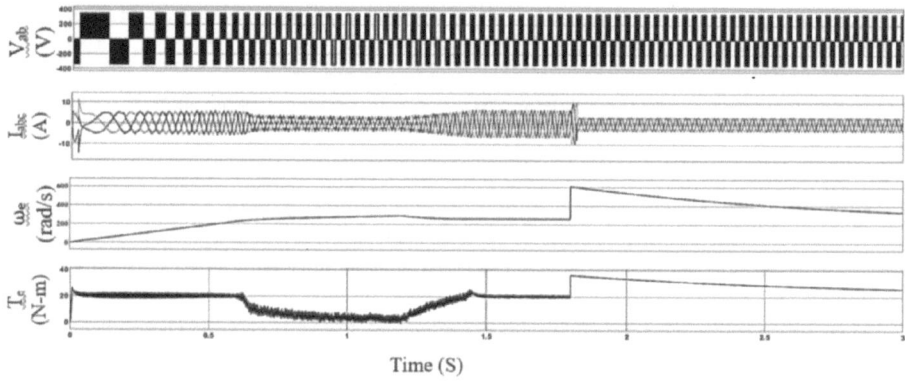

Time (S)

FIGURE 7.19   Performance of IMPD with the reduction in insolation at 1.8 seconds.

REFERENCES

[1] Shanshan, G., Yueshi, G.: 'A High frequency high voltage gain modified SEPIC with integrated inductors', IEEE transactions on Industry Applications, 2019, 55 (6), pp. 7481–7490.

[2] Mahamoodreja, E. A., Hossein, S.: 'A voltage gain DC–DC converter based on three-winding coupled inductor and voltage multiplier cell', IEEE transactions on Power Electronics, 2020, 35 (5), pp. 4558–4567.

[3] Igor, R. D. O., Aniel, S. D. M., Fernando, L. T.: 'Single-switch integrated DC–DC converter for high voltage step down applications', IET Power Electronics, 2019, 12 (8), pp. 1880–1890.

[4] Antonio, M. S. S. A., Luciano, S.: 'synthesis and comparative analysis of very high step-up DC–DC converters adopting coupled inductor and voltage multiplier cells', IEEE Transactions on Power Electronics, 2018, 33 (7) pp. 5880–5897.

[5] Zhang, Z., Thomsen, O. C., Andersen, M. A. E., Nielsen, H. R.: 'Dual input isolated full-bridge boost DC–DC converter based on the distributed transformers', IET Power Electronics, 2012, 5 (7), pp. 1074–1083

[6] Changchien, S. K., Liang, T. J., Chen., J. F., Yang, L. S.: 'Step-up DC–DC converter by coupled inductor and voltage-lift technique', IET Power Electronics, 2010, 3 (3), pp. 369–378.

[7] Prudente, M., Luciano, L., Pfitscher, Gustavo E., Romaneli., E. F., Gules, R.: 'Voltage multiplier cells applied to non-isolated DC–DC converters', IEEE Transactions on Power Electronics, 2008, 23 (2), pp. 871–887.

[8] Sara, M. F., Mehran, S.: 'High step-down/step-up interleaved bi-directional DC–DC converter with low voltage stress on switches', IET Power Electronics, 2019, 13 (1), pp. 104–115.

[9] Park, Ki-Bum, Moon, Gun-Woo, Youn, Myung-Joong.: 'Non-isolated high step-up boost converter integrated with SEPIC converter', IEEE Transactions on Power Electronics, 2010, 25 (9), pp. 2266–2275.

[10] Dongyan, Z., Pietkiewicz, A., Cuk, S.: 'A three-switch high voltage converter', IEEE Transactions on Power Electronics, 1999, 14 (1), pp. 177–183.

[11] Tofoli, F.L., Perrira, D. de C., Paulo, W. J. de., Junior, D. de S. O.: 'Survey on non-isolated high voltage step-up DC–DC topologies based on the boost converter', IET Power Electronics, 2015, ISSN:1755–4535, pp. 1–14.

[12] Ismail, E.H., Alsaffar, M.A., Sabzali, A.J.: 'High conversion ratio DC–DC converters with reduced switch stress', IEEE Transactions on Circuit and Systems I, 2008, 55 (7), pp. 2139–2151.

[13] Ranjana, M.S.B., Kulkarni, Rishi M., Kohkade, Anita, Chandodkar, Pooja.: 'Non-isolated switched inductor SEPIC Converter Topologies for Photovoltaic boost application', IEEE International Conference on Circuit, Power and Computing Technology (ICCPCT), 2016, pp. 1–6.

[14] Suryoatmojo, H., Dilianto, I., Suwito, R. M., Setijadi, E., Riawan, D.C.: 'Design and anaysis of high gain modified SEPIC converter for Photovoltaic applications', IEEE International Conference on Innovative Research and Development (ICIRD), 2018, pp. 1–6.

[15] Maroti, Pandav K., Sanjeevikumar, Padmanaban, Frede, Blaabjerg, Dan, Ionel.: 'Novel immense configurations of boost converter for renewable energy application', The Journal of Engineering, 2019, 18, pp. 4884–4889.

[16] Gao, S., Wang, Y., Xu, D.: 'Modified SEPIC converter with high voltage gain and ZVS characteristics', IEEE Transactions on Circuits and Systems II: Express Briefs, 2019, 66 (11), pp. 1860–1864.

[17] Neha, B., Hussain, I., Singh, B.: 'Vector-based synchronization method for grid integration of solar PV-battery system', The IEEE Transactions on Industrial Informatics, Jun. 2019, 15 (9), pp. 4923–4933.

[18] Kumar, Nishant, Singh, B., Panigrahi, B. K., Chakraborty, C., Suryawanshi, H. M., Verma, V.: 'Integration of solar PV with low-voltage weak grid system: Using normalized Laplacian kernel adaptive Kalman filter and learning based InC algorithm', IEEE Transactions on Power Electronics, Nov. 2019, 34 (11), pp. 10746–10758.

[19] Singh, B., Kumar, Shailendra: 'Grid integration of 3P4W solar PV system using M-LWDF-based control technique', IET Renewable Power Generation, 2016, 11 (8), pp. 1174–1181.

[20] Pradhan, S., Hussain, I., Singh, B., Panigrahi, B. K.: Performance improvement of grid-integrated solar PV system using DNLMS control algorithm', IEEE Transactions on Industry Applications, Aug. 2018, 55 (1), pp. 78–91.

[21] Shukl, P., Singh, B.: 'Grid integration of three-phase single-stage PV system using adaptive laguerre filter based control algorithm under nonideal distribution system', IEEE Transactions on Industry Applications, Jul. 2019, 55 (6), pp. 6193–6202.

[22] Patra, S., Kishor, N., Mohanty, S. R., Ray, P. K.: 'Power quality assessment in 3-Φ grid connected PV system with single and dual stage circuits', International Journal of Electrical Power & Energy Systems, 2016, 75, pp. 275–288.

[23] Tomar, A., Mishra, S., Bhende, C. N.: 'AOMH—MISO based PV—VCI irrigation system using ASCIM pump', IEEE Transactions on Industry Applications, May, 2018, 54 (5), pp. 4813–4824.

[24] Vitorino, M. A., de Rossiter Corrêa, M. B., Jacobina, C. B., Lima, A. M. N.: 'An effective induction motor control for photovoltaic pumping', IEEE Transactions on Industrial Electronics, Jun. 2010, 58 (4), pp. 1162–1170.

[25] Caracas, J. V. M., de Carvalho Farias, G., Teixeira, L. F. M., de Souza Ribeiro, L. A.: 'Implementation of a high-efficiency, high-lifetime, and low-cost converter for an autonomous photovoltaic water pumping system', IEEE Transactions on Industry Applications, Jun. 2013, 50 (1), pp. 631–641.

[26] Mudlapur, A., Ramana, V. V., Damodaran, R. V., Balasubramanian, V., Mishra, S.: 'Effect of partial shading on PV fed induction motor water pumping systems', IEEE Transactions on Energy Conversion, Oct. 2018, 34 (1), pp. 530–539.

[27] Muljadi, E.: 'PV water pumping with a peak-power tracker using a simple six-step square-wave inverter', IEEE Transactions on Industry Applications, May/Jun. 1997, 33 (3), pp. 714–721.

[28] Correa, O. P., Seleme, S. I., Silva, S. R.: 'Efficiency optimization in stand-alone photo-voltaic pumping system', Renewable Energy, 2012, 41, pp. 220–226.

[29] Abella, M. A., Lorenzo, E., Chenlo, F.: 'PV water pumping systems based on standard frequency converters', Photovoltaics: Research and Applications, 2003, 11, pp. 179–191.

APPENDICES

Appendix A1: Arbitrary Constants of Configuration 1

$$r_e = 2r_{l1} + r_{c1} + r_{d2} \tag{A1.1}$$

$$\tau_{cd2} = (r_{c2} + r_{d2})C_2 \tag{A1.2}$$

$$\tau_{R4} = (r_{c4} + R) . C_4 \tag{A1.3}$$

$$\tau_4 = r_{c4} . C_4 \tag{A1.4}$$

$$R_{eq} = R . \frac{\tau_4}{\tau_{R4}} = R \| {}^{r_{c4}} \tag{A1.5}$$

$$\tau_{c2} = (r_{c2} + r_{c3} + r_{d2} + r_{d3} + R_{eq})C_2 \tag{A1.6}$$

$$\tau_1 = C_1 (2r_{d1} + r_{c1}) \tag{A1.7}$$

$$\tau_{11} = C_1 (r_{d1} + r_{c1}) \tag{A1.8}$$

$$\tau_2 = \frac{r_{L2} + r_{c2} + r_{c3}}{L_2} \tag{A1.9}$$

$$\tau_{c4} = C_4 (r_{c4} + R) \tag{A1.10}$$

$$r_{3d3eq} = r_{c3} + r_{d3} + R_{eq} \tag{A1.11}$$

$$a_{11}' = \frac{r_{d1}^2 C_1}{L_1 \tau_1} - \frac{r_{d1} + r_{L1}}{L_1} \tag{A1.12}$$

$$a'_{12} = \frac{r_{d1}^2 C_1}{L_1 \tau_1}$$ (A1.13)

$$a'_{13} = \frac{r_{d1} C_1}{L_1 \tau_1}$$ (A1.14)

$$a'_{21} = \frac{r_{d1}}{L_1}\left[1 - \frac{\tau_{11}}{\tau_1}\right]$$ (A1.15)

$$a'_{22} = -\left[\frac{r_{L1}}{L_1} - \frac{r_{d1}}{L_1}\frac{\tau_{11}}{\tau_1}\right]$$ (A1.16)

$$a'_{23} = \frac{1}{L_1}\left[1 - \frac{\tau_{11}}{\tau_1}\right].$$ (A1.17)

$$a'_{31} = -\frac{r_{d1}}{\tau_1}$$ (A1.18)

$$a'_{32} = -\frac{r_{d1}}{\tau_1}$$ (A1.19)

$$a'_{33} = -\frac{1}{\tau_1}$$ (A1.20)

$$a'_{44} = -\tau_2$$ (A1.21)

$$a'_{45} = a'_{46} = \frac{-1}{L_2}$$ (A1.22)

$$a'_{54} = \frac{1}{C_2}$$ (A1.23)

$$a'_{64} = -\frac{1}{C_2}$$ (A1.24)

$$a'_{77} = \frac{-1}{\tau_{c4}}$$ (A1.25)

$$b'_{11} = \frac{1}{L_1} - \frac{r_{d1}C_1}{L_1\tau_1} \tag{A1.26}$$

$$b'_{12} = \frac{2r_{d1}C_1}{L_1\tau_1} - \frac{1}{L_1} \tag{A1.27}$$

$$b'_{21} = \frac{\tau_{11}}{L_1\tau_1} \tag{A1.28}$$

$$b'_{22} = \frac{1}{L_1}\left[1 - \frac{2\tau_{11}}{\tau_1}\right] \tag{A1.29}$$

$$C'_7 = \frac{RC_4}{\tau_{c4}} \tag{A1.30}$$

$$a_{11} = \frac{1}{2L_1}\left(\frac{C_2\left(r_{d2} + r_{c3}\right)^2}{\tau_{c2}} - \left(r_{c3} + r_e\right)\right) \tag{A1.31}$$

$$a_{13} = \frac{1}{2L_1} \tag{A1.32}$$

$$a_{14} = \frac{1}{2L_1}\left(\frac{\left(r_{d2} + r_{c3}\right)\left(r_{3d3eq}\right)C_2}{\tau_{c2}} - \left(r_{c3}\right)\right) \tag{A1.33}$$

$$a_{15} = \frac{1}{2L_1}\left(\frac{\left(r_{d2} + r_{c3}\right)C_2}{\tau_{c2}}\left[\frac{\tau_4}{\tau_{R4}} - 1\right]\right) \tag{A1.34}$$

$$a_{16} = \frac{1}{2L_1}\left(\frac{\left(r_{d2} + r_{c3}\right)C_2}{\tau_{c2}} - 1\right) \tag{A1.35}$$

$$a_{17} = \left(\frac{\left(r_{d2} + r_{c3}\right)C_2}{2L_1\tau_{c2}}\left[\frac{\tau_4}{\tau_{R4}} - 1\right]\right) \tag{A1.36}$$

$$a_{31} = \frac{1}{c_1} \tag{A1.37}$$

$$a_{41} = \frac{\tau_{cd2}}{\tau_{c2}}\frac{\left(r_{d2} + r_{c3}\right)}{L_2} - \frac{r_{d2}}{L_2} \tag{A1.38}$$

$$a_{44} = \frac{\tau_{cd2}}{\tau_{c2}} \frac{\left(r_{d3} + r_{c3} + R_{eq}\right)}{L_2} + \frac{r_{12}}{L_2} \qquad (A1.39)$$

$$a_{45} = \frac{1}{L_2}\left[1 - \frac{\tau_{cd2}}{\tau_{c2}}\right] \qquad (A1.40)$$

$$a_{46} = \frac{1}{L_2}\left[\frac{\tau_{cd2}}{\tau_{c2}}\right] \qquad (A1.41)$$

$$a_{47} = \frac{1}{L_2}\frac{\tau_{cd2}}{\tau_{c2}}\left[\frac{\tau_4}{\tau_{R4}} - 1\right] \qquad (A1.42)$$

$$a_{51} = -\left(\frac{\left(r_{d2} + r_{c3}\right)}{\tau_{c2}}\right) \qquad (A1.43)$$

$$a_{54} = -\left(\frac{\left(r_{c3} + r_{d3} + R_{eq}\right)}{\tau_{c2}}\right) \qquad (A1.44)$$

$$a_{55} = \frac{1}{\tau_{c2}} \qquad (A1.45)$$

$$a_{56} = -\frac{1}{\tau_{c2}} \qquad (A1.46)$$

$$a_{57} = \frac{1}{\tau_{c2}}\left[1 - \frac{\tau_4}{\tau_{R4}}\right] \qquad (A1.47)$$

$$a_{61} = -\frac{1}{C_3}\left(1 - \frac{\left(r_{d2} + r_{c3}\right)C_2}{\tau_{c2}}\right) \qquad (A1.48)$$

$$a_{64} = -\frac{1}{C_3}\left(\frac{\left(r_{c3} + r_{d3} + R_{eq}\right)C_2}{\tau_{c2}} - 1\right) \qquad (A1.49)$$

$$a_{65} = -\frac{C_2}{C_3}\left[\frac{1}{\tau_{c2}}\right] \qquad (A1.50)$$

$$a_{66} = \frac{C_2}{C_3}\left[\frac{1}{\tau_{c2}}\right] \qquad (A1.51)$$

$$a_{67} = \frac{C_2}{C_3}\frac{1}{\tau_{c2}}\left[\frac{\tau_4}{\tau_{R4}} - 1\right]$$

(A1.52)

$$a_{71} = \frac{RC_2}{\tau_{R4}}\left(\frac{(r_{d2} + r_{c3})}{\tau_{c2}}\right)$$

(A1.53)

$$a_{74} = \frac{R}{\tau_{R4}}\left(\frac{(r_{c3} + r_{d3} + R_{eq})C_2}{\tau_{c2}} - 1\right)$$

(A1.54)

$$a_{75} = -\frac{RC_2}{\tau_{R4}\tau_{c2}}$$

(A1.55)

$$a_{76} = \frac{RC_2}{\tau_{R4}\tau_{c2}}$$

(A1.56)

$$a_{77} = \frac{1}{\tau_{R4}}\left\{\frac{RC_2}{\tau_{c2}}\left[\frac{\tau_4}{\tau_{R4}} - 1\right] - 1\right\}$$

(A1.57)

$$b_{11} = \frac{1}{2L_1}$$

(A1.58)

$$b_{13} = \frac{1}{2L_1}\left(\frac{C_2(r_{d2} + r_{c3})}{\tau_{c2}} - 1\right)$$

(A1.59)

$$b_{14} = -\left(\frac{C_2(r_{d2} + r_{c3})}{2L_1\tau_{c2}}\right)$$

(A1.60)

$$b_{43} = \frac{1}{L_2}\left[\frac{\tau_{cd2}}{\tau_{c2}} - 1\right]$$

(A1.61)

$$b_{44} = -\frac{1}{L_2}\frac{\tau_{cd2}}{\tau_{c2}}$$

(A1.62)

$$b_{53} = -\frac{1}{\tau_{c2}}$$

(A1.63)

$$b_{54} = \frac{1}{\tau_{c2}}$$

(A1.64)

$$b_{63} = \frac{C_2}{C_3}\left[\frac{1}{\tau_{c2}}\right] \tag{A1.65}$$

$$b_{64} = -\frac{C_2}{C_3}\left[\frac{1}{\tau_{c2}}\right] \tag{A1.66}$$

$$b_{73} = \frac{RC_2}{\tau_{R4}\tau_{c2}} \tag{A1.67}$$

$$b_{74} = -\frac{RC_2}{\tau_{R4}\tau_{c2}} \tag{A1.68}$$

$$c_{11} = \frac{RC_2}{\tau_{c2}}\left(r_{d2} + r_{c3}\right)\left(1 - \frac{RC_4}{\tau_{R4}}\right) \tag{A1.69}$$

$$c_{14} = R\left(1 - \frac{RC_4}{\tau_{R4}}\right)\left[\frac{\left(r_{c3} + r_{d3} + R_{eq}\right)C_2}{\tau_{c2}} - 1\right] \tag{A1.70}$$

$$c_{15} = \frac{RC_2}{\tau_{c2}}\left(\frac{RC_4}{\tau_{R4}} - 1\right) \tag{A1.71}$$

$$c_{16} = \frac{RC_2}{\tau_{c2}}\left(1 - \frac{RC_4}{\tau_{R4}}\right) \tag{A1.72}$$

$$c_{17} = R\left[\frac{C_4}{\tau_{R4}} + \frac{C_2}{\tau_{c2}}\left[1 - \frac{\tau_4}{\tau_{R4}}\right]\left(\frac{RC_4}{\tau_{R4}} - 1\right)\right] \tag{A1.73}$$

$$d_{13} = \frac{RC_2}{\tau_{c2}}\left(1 - \frac{RC_4}{\tau_{R4}}\right) \tag{A1.74}$$

$$d_{14} = \frac{RC_2}{\tau_{c2}}\left(\frac{RC_4}{\tau_{R4}} - 1\right) \tag{A1.75}$$

APPENDIX A2: ARBITRARY CONSTANTS OF CONFIGURATION 2

$$r_e = 2r_{l2} + r_{c2} \tag{A2.1}$$

$$r_1 = r_{c1} + r_{c3} + r_{d2} \tag{A2.2}$$

$$r_2 = r_{c1} + r_{c2} + r_{c3} + 2r_{d2} \tag{A2.3}$$

$$r_{2d2} = r_{c2} + r_{d2} \tag{A2.4}$$

$$\tau_{21} = \tau_{c1r} = r_2 C_1 \tag{A2.5}$$

$$\tau_{23} = r_2 C_3 \tag{A2.6}$$

$$\tau_{c2r} = r_2 C_2 \tag{A2.7}$$

$$\tau_{c4} = C_4 \left(r_{c4} + R \right) \tag{A2.8}$$

$$R' = R \parallel r_{c4} = \frac{R r_{c4}}{R + r_{c4}} \tag{A2.9}$$

$$r_{1d1} = r_{c1} + r_{d1} \tag{A2.10}$$

$$r_{3d1} = r_{c3} + r_{d1} \tag{A2.11}$$

$$r'_{3d3} = r_{c3} + r_{d3} + R' \tag{A2.12}$$

$$r_3 = r_{l1} + r_{c3} + r_{d1} \tag{A2.13}$$

$$r_4 = r_{c1} + r_{c3} + r_{d1} + r_{d3} + R' \tag{A2.14}$$

$$\tau_4 = r_{c4} C_4 \tag{A2.15}$$

$$\tau_{4c1} = r_4 C_1 \tag{A2.16}$$

$$a'_{11} = -\frac{r_{l1}}{L_1} \tag{A2.17}$$

$$a'_{22} = \frac{-1}{L_2} \left[r_{l2} + r_{2d2} \frac{r_1}{r_2} \right] \tag{A2.18}$$

$$a'_{23} = \frac{1}{L_2} \left[r_{d2} - r_{2d2} \frac{r_1}{r_2} \right] \tag{A2.19}$$

$$a'_{24} = \frac{-1}{L_2} \left[\frac{r_{2d2}}{r_2} \right] \tag{A2.20}$$

$$a'_{25} = \frac{1}{L_2} \left[\frac{r_{2d2}}{r_2} - 1 \right] \tag{A2.21}$$

$$a'_{26} = \frac{-1}{L_2} \left[\frac{r_{2d2}}{r_2} \right] \tag{A2.22}$$

$$a'_{32} = \frac{1}{L_2}\left[r_{d2} - r_{2d2}\frac{r_1}{r_2} \right] \qquad \text{(A2.23)}$$

$$a'_{33} = \frac{-1}{L_2}\left[r_{12} + r_{2d2}\frac{r_1}{r_2} \right] \qquad \text{(A2.24)}$$

$$a'_{34} = \frac{-1}{L_2}\left[\frac{r_{2d2}}{r_2} \right] \qquad \text{(A2.25)}$$

$$a'_{35} = \frac{1}{L_2}\left[\frac{r_{2d2}}{r_2} - 1 \right] \qquad \text{(A2.26)}$$

$$a'_{36} = \frac{-1}{L_2}\left[\frac{r_{2d2}}{r_2} \right] \qquad \text{(A2.27)}$$

$$a'_{42} = \frac{r_{2d2}}{\tau_{21}} \qquad \text{(A2.28)}$$

$$a'_{43} = \frac{r_{2d2}}{\tau_{21}} \qquad \text{(A2.29)}$$

$$a'_{44} = \frac{1}{\tau_{21}} \qquad \text{(A2.30)}$$

$$a'_{45} = \frac{-1}{\tau_{21}} \qquad \text{(A2.31)}$$

$$a'_{46} = \frac{1}{\tau_{21}} \qquad \text{(A2.32)}$$

$$a'_{52} = \frac{r_1}{\tau_{c2r}} \qquad \text{(A2.33)}$$

$$a'_{53} = \frac{r_1}{\tau_{c2r}} \qquad \text{(A2.34)}$$

$$a'_{54} = \frac{1}{\tau_{c2r}} \qquad \text{(A2.35)}$$

$$a'_{55} = \frac{-1}{\tau_{c2r}} \qquad \text{(A2.36)}$$

$$a'_{56} = \frac{1}{\tau_{c2r}} \qquad\qquad (A2.37)$$

$$a'_{62} = \frac{r_{2d2}}{\tau_{23}} \qquad\qquad (A2.38)$$

$$a'_{63} = \frac{r_{2d2}}{\tau_{23}} \qquad\qquad (A2.39)$$

$$a'_{64} = \frac{1}{\tau_{23}} \qquad\qquad (A2.40)$$

$$a'_{65} = \frac{-1}{\tau_{23}} \qquad\qquad (A2.41)$$

$$a'_{66} = \frac{1}{\tau_{23}} \qquad\qquad (A2.42)$$

$$a'_{77} = \frac{1}{\tau_{c4}} \qquad\qquad (A2.43)$$

$$b'_{11} = \frac{1}{L_1} \qquad\qquad (A2.44)$$

$$b'_{23} = \frac{1}{L_2}\left[\frac{r_{2d2}}{r_2} - 1\right] \qquad\qquad (A2.45)$$

$$b'_{33} = \frac{1}{L_2}\left[\frac{r_{2d2}}{r_2} - 1\right] \qquad\qquad (A2.46)$$

$$b'_{43} = \frac{2}{\tau_{21}} \qquad\qquad (A2.47)$$

$$b'_{53} = \frac{-2}{\tau_{c2r}} \qquad\qquad (A2.48)$$

$$b'_{63} = \frac{2}{\tau_{23}} \qquad\qquad (A2.49)$$

$$c'_7 = \frac{R'}{r_{c4}} \qquad\qquad (A2.50)$$

$$a_{11} = \frac{1}{L_1}\left(r_3 - \frac{r_{3d1}^2}{r_4}\right) \tag{A2.51}$$

$$a_{12} = \frac{1}{L_1}\left(r_{c3} - \frac{r_{3d1}r_{3d3}'}{r_4}\right) \tag{A2.52}$$

$$a_{14} = \frac{1}{L_1}\left(\frac{r_{3d1}}{r_4}\right) \tag{A2.53}$$

$$a_{16} = \frac{1}{L_1}\left(1 - \frac{r_{3d1}}{r_4}\right) \tag{A2.54}$$

$$a_{17} = \frac{1}{L_1}\frac{r_{3d1}}{r_4}\left(1 - \frac{R'}{R}\right) \tag{A2.55}$$

$$a_{21} = \frac{1}{2L_2}\left(\frac{r_{1d1}r_{3d1}}{r_4} - r_{d1}\right) \tag{A2.56}$$

$$a_{22} = \frac{1}{2L_2}\left(r_e - \frac{r_{1d1}r_{3d3}'}{r_4}\right) \tag{A2.57}$$

$$a_{24} = \frac{1}{2L_2}\left(1 - \frac{r_{1d1}}{r_4}\right) \tag{A2.58}$$

$$a_{25} = \frac{1}{2L_2} \tag{A2.59}$$

$$a_{26} = \frac{1}{2L_2}\frac{r_{1d1}}{r_4} \tag{A2.60}$$

$$a_{27} = \frac{1}{2L_2}\frac{r_{1d1}}{r_4}\left(\frac{R'}{R} - 1\right) \tag{A2.61}$$

$$a_{41} = -\frac{r_{3d1}}{C_1 r_4} \tag{A2.62}$$

$$a_{42} = -\frac{r_{3d3}'}{C_1 r_4} \tag{A2.63}$$

$$a_{44} = \frac{1}{C_1 r_4} \tag{A2.64}$$

$$a_{46} = -\frac{1}{C_1 r_4} \tag{A2.65}$$

$$a_{47} = \frac{1}{C_1 r_4}\left(1 - \frac{R'}{R}\right) \tag{A2.66}$$

$$a_{52} = \frac{-1}{C_2} \tag{A2.67}$$

$$a_{61} = \frac{1}{C_3}\left(\frac{r_{3d1}}{r_4} - 1\right) \tag{A2.68}$$

$$a_{62} = \frac{1}{C_3}\left(\frac{r'_{3d3}}{r_4} - 1\right) \tag{A2.69}$$

$$a_{64} = \frac{-1}{r_4 C_3} \tag{A2.70}$$

$$a_{66} = \frac{1}{r_4 C_3} \tag{A2.71}$$

$$a_{67} = \frac{-1}{C_3 r_4}\left(1 - \frac{R'}{R}\right) \tag{A2.72}$$

$$a_{71} = \frac{R' r_{3d1}}{\tau_4 r_4} \tag{A2.73}$$

$$a_{72} = \frac{R'}{\tau_4}\left(\frac{r'_{3d3}}{r_4} - 1\right) \tag{A2.74}$$

$$a_{74} = \frac{-R'}{r_4 \tau_4} \tag{A2.75}$$

$$a_{76} = \frac{R'}{r_4 \tau_4} \tag{A2.76}$$

$$a_{77} = \frac{R'}{\tau_4}\left(\frac{1}{r_4}\left(\frac{R'}{R} - 1\right) - \frac{1}{R}\right) \tag{A2.77}$$

$$b_{11} = \frac{-1}{L_1} \tag{A2.78}$$

$$b_{12} = \frac{1}{L_1}\left(1 - \frac{r_{3d1}}{r_4}\right)$$

(A2.79)

$$b_{14} = \frac{1}{L_1}\frac{r_{3d1}}{r_4}$$

(A2.80)

$$b_{22} = \frac{1}{L_2}\left(\frac{r_{1d1}}{r_4} - 1\right)$$

(A2.81)

$$b_{24} = \frac{-1}{L_2}\left(\frac{r_{1d1}}{r_4}\right)$$

(A2.82)

$$b_{42} = \frac{-1}{C_1 r_4}$$

(A2.83)

$$b_{44} = \frac{1}{C_1 r_4}$$

(A2.84)

$$b_{62} = \frac{1}{r_4 C_3}$$

(A2.85)

$$b_{64} = \frac{-1}{r_4 C_3}$$

(A2.86)

$$b_{72} = \frac{R'}{r_4 \tau_4}$$

(A2.87)

$$b_{74} = \frac{-R'}{r_4 \tau_4}$$

(A2.88)

$$C_{11} = R\frac{r_{3d1}}{r_4}\left(1 - \frac{R'}{r_{c4}}\right)$$

(A2.89)

$$C_{12} = R\left(\frac{r'_{3d3}}{r_4} - 1\right)\left(1 - \frac{R'}{r_{c4}}\right)$$

(A2.90)

$$C_{14} = \frac{R}{r_4}\left(\frac{R'}{r_{c4}} - 1\right)$$

(A2.91)

$$C_{16} = \frac{R}{r_4}\left(1 - \frac{R'}{r_{c4}}\right)$$

(A2.92)

$$C_{17} = \frac{R'}{r_{c4}} + \frac{R}{r_4}\left(1 - \frac{R'}{r_{c4}}\right)\left(\frac{R'}{R} - 1\right) \tag{A2.93}$$

$$D_{12} = \frac{R}{r_4}\left(1 - \frac{R'}{r_{c4}}\right) \tag{A2.94}$$

$$D_{14} = \frac{R}{r_4}\left(\frac{R'}{r_{c4}} - 1\right) \tag{A2.95}$$

APPENDIX A3: ARBITRARY CONSTANTS OF CONFIGURATION 3

$$\tau_{c1d1} = (r_{c1} + 2r_{d1})\ C_1 \tag{A3.1}$$

$$r_1 = r_{c2} + r_{c4} + r_{d2} \tag{A3.2}$$

$$r_2 = r_{c2} + r_{c3} + r_{c4} + 2r_{d2} \tag{A3.3}$$

$$r_{3d2} = r_{c3} + r_{d2} \tag{A3.4}$$

$$\tau_2 = C_2\, r_2 \tag{A3.5}$$

$$\tau_{c3r} = C_3\, r_2 \tag{A3.6}$$

$$\tau_4 = C_4\, r_2 \tag{A3.7}$$

$$\tau_{c5} = (R + r_{c5})C_5 \tag{A3.8}$$

$$R' = R \,\|\, r_{c5} = \frac{Rr_{c5}}{R + r_{c5}} \tag{A3.9}$$

$$r_3 = r_{c2} + r_{c4} + r_{d3} + r_{d4} + R' \tag{A3.10}$$

$$r_4' = r_{c4} + r_{d4} + R' \tag{A3.11}$$

$$r_4 = r_{c1} + 2r_{l1} + r_{c4} + r_{d3} \tag{A3.12}$$

$$r_{43} = r_{c4} + r_{d3} \tag{A3.13}$$

$$r_{23} = r_{c2} + r_{d3} \tag{A3.14}$$

$$r_{e2} = r_{c3} + 2r_{l2} \tag{A3.15}$$

$$\tau_{32} = r_3 C_2 \tag{A3.16}$$

$$a'_{11} = \frac{1}{L_{1a}} \left(\frac{r_{d1}^2 C_1}{\tau_{c1d1}} - (r_{l1a} + r_{d1}) \right) \tag{A3.17}$$

$$a'_{12} = \frac{1}{L_{1a}} \left(\frac{r_{d1}^2 C_1}{\tau_{c1d1}} \right) \tag{A3.18}$$

$$a'_{13} = \frac{1}{L_{1a}} \left(\frac{r_{d1} C_1}{\tau_{c1d1}} \right) \tag{A3.19}$$

$$a'_{21} = \frac{1}{L_{1b}} \left(\frac{r_{d1}^2 C_1}{\tau_{c1d1}} \right) \tag{A3.20}$$

$$a'_{23} = \frac{1}{L_{1b}} \left(\frac{r_{d1} C_1}{\tau_{c1d1}} \right) \tag{A3.21}$$

$$a'_{22} = \frac{1}{L_{1b}} \left(\frac{r_{d1}^2 C_1}{\tau_{c1d1}} - (r_{l1b} + r_{d1}) \right) \tag{A3.22}$$

$$a'_{31} = \frac{-r_{d1} C_1}{\tau_{c1d1}} \tag{A3.23}$$

$$a'_{32} = \frac{-r_{d1} C_1}{\tau_{c1d1}} \tag{A3.24}$$

$$a'_{33} = \frac{-1}{\tau_{c1d1}} \tag{A3.25}$$

$$a'_{44} = \frac{-1}{L_{2a}} \left(r_{l2a} + r_{3d2} \frac{r_1}{r_2} \right) \tag{A3.26}$$

$$a'_{45} = \frac{1}{L_{2a}} \left(r_{d2} - r_{3d2} \frac{r_1}{r_2} \right) \tag{A3.27}$$

$$a'_{46} = \frac{-1}{L_{2a}} \left(\frac{r_{3d2}}{r_2} \right) \tag{A3.28}$$

$$a'_{47} = \frac{1}{L_{2a}} \left(\frac{r_{3d2}}{r_2} - 1 \right) \tag{A3.29}$$

$$a'_{48} = \frac{-1}{L_{2a}} \left(\frac{r_{3d2}}{r_2} \right) \tag{A3.30}$$

$$a'_{54} = \frac{1}{L_{2b}}\left(r_{d2} - r_{3d2}\frac{r_1}{r_2}\right)$$ (A3.31)

$$a'_{55} = \frac{-1}{L_{2b}}\left(r_{12b} + r_{3d2}\frac{r_1}{r_2}\right)$$ (A3.32)

$$a'_{56} = \frac{-1}{L_{2b}}\left(\frac{r_{3d2}}{r_2}\right)$$ (A3.33)

$$a'_{57} = \frac{1}{L_{2b}}\left(\frac{r_{3d2}}{r_2} - 1\right)$$ (A3.34)

$$a'_{58} = \frac{-1}{L_{2b}}\left(\frac{r_{3d2}}{r_2}\right)$$ (A3.35)

$$a'_{64} = \frac{-r_{3d2}}{\tau_2}$$ (A3.36)

$$a'_{65} = \frac{r_{3d2}}{\tau_2}$$ (A3.37)

$$a'_{66} = -\frac{1}{\tau_2}$$ (A3.38)

$$a'_{67} = \frac{1}{\tau_2}$$ (A3.39)

$$a'_{68} = \frac{1}{\tau_2}$$ (A3.40)

$$a'_{74} = \frac{r_1}{\tau_{c3r}}$$ (A3.41)

$$a'_{75} = \frac{r_1}{\tau_{c3r}}$$ (A3.42)

$$a'_{76} = \frac{1}{\tau_{c3r}}$$ (A3.43)

$$a'_{77} - \frac{1}{\tau_{c3r}}$$ (A3.44)

$$a'_{78} = \frac{1}{\tau_{c3r}} \tag{A3.45}$$

$$a'_{84} = \frac{r_{3d2}}{\tau_4} \tag{A3.46}$$

$$a'_{85} = \frac{r_{3d2}}{\tau_4} \tag{A3.47}$$

$$a'_{86} = \frac{-1}{\tau_4} \tag{A3.48}$$

$$a'_{87} = \frac{1}{\tau_4} \tag{A3.49}$$

$$a'_{88} = \frac{-1}{\tau_4} \tag{A3.50}$$

$$a'_{99} = \frac{1}{\tau_{c5}} \tag{A3.51}$$

$$b'_{11} = \frac{1}{L_{1a}}\left(1 - \frac{r_{d1}C_1}{\tau_{c1d1}}\right) \tag{A3.52}$$

$$b'_{12} = \frac{1}{L_{1a}}\left(\frac{2r_{d1}C_1}{\tau_{c1d1}} - 1\right) \tag{A3.53}$$

$$b'_{21} = \frac{1}{L_{1b}}\left(1 - \frac{r_{d1}C_1}{\tau_{c1d1}}\right) \tag{A3.54}$$

$$b'_{22} = \frac{1}{L_{1a}}\left(\frac{2r_{d1}C_1}{\tau_{c1d1}} - 1\right) \tag{A3.55}$$

$$b'_{31} = \frac{1}{\tau_{c1d1}} \tag{A3.56}$$

$$b'_{32} = \frac{-2}{\tau_{c1d1}} \tag{A3.57}$$

$$b'_{43} = \frac{1}{L_{2a}}\left(\frac{2r_{3d2}}{r_2} - 1\right) \tag{A3.58}$$

$$b'_{53} = \frac{1}{L_{2b}}\left(\frac{2r_{3d2}}{r_2} - 1\right) \tag{A3.59}$$

$$b'_{63} = \frac{2}{\tau_2} \tag{A3.60}$$

$$b'_{73} = \frac{-2}{\tau_{c3r}} \tag{A3.61}$$

$$b'_{83} = \frac{2}{\tau_4} \tag{A3.62}$$

$$c'_9 = \frac{R'}{r_{c5}} \tag{A3.63}$$

$$a_{11} = \frac{1}{2L_1}\left(r_4 - \frac{r_{43}^2}{r_3}\right) \tag{A3.64}$$

$$a_{13} = \frac{-1}{2L_1} \tag{A3.65}$$

$$a_{14} = \frac{1}{2L_1}\left(r_{c4} - \frac{r_{43}r'_4}{r_3}\right) \tag{A3.66}$$

$$a_{16} = \frac{1}{2L_1}\left(\frac{r_{43}}{r_3}\right) \tag{A3.67}$$

$$a_{18} = \frac{1}{2L_1}\left(1 - \frac{r_{43}}{r_3}\right) \tag{A3.68}$$

$$a_{19} = \frac{1}{2L_1}\left(\frac{r_{43}R'}{r_{c5}r_3}\right) \tag{A3.69}$$

$$a_{31} = \frac{-1}{C_1} \tag{A3.70}$$

$$a_{41} = \frac{1}{2L_2}\left(\frac{r_{43}r_{23}}{r_3} - r_{d3}\right) \tag{A3.71}$$

$$a_{44} = \frac{1}{2L_2}\left(\frac{r_{43}r'_4}{r_3} + r_{e2}\right) \tag{A3.72}$$

$$a_{46} = \frac{1}{2L_2}\left(1 - \frac{r_{23}}{r_3}\right) \tag{A3.73}$$

$$a_{47} = \frac{1}{2L_2} \tag{A3.74}$$

$$a_{48} = \frac{1}{2L_2}\frac{r_{23}}{r_3} \tag{A3.75}$$

$$a_{49} = \frac{-1}{2L_2}\frac{R'r_{23}}{r_3 r_{c5}} \tag{A3.76}$$

$$a_{61} = \frac{-r_{43}}{\tau_{32}} \tag{A3.77}$$

$$a_{64} = \frac{-r_4'}{\tau_{32}} \tag{A3.78}$$

$$a_{66} = \frac{1}{\tau_{32}} \tag{A3.79}$$

$$a_{68} = \frac{-1}{\tau_{32}} \tag{A3.80}$$

$$a_{69} = \frac{R'}{\tau_{32}r_{c5}} \tag{A3.81}$$

$$a_{74} = \frac{-1}{C_3} \tag{A3.82}$$

$$a_{81} = \frac{1}{C_4}\left(\frac{r_{43}}{r_3} - 1\right) \tag{A3.83}$$

$$a_{84} = \frac{1}{C_4}\left(\frac{r_4'}{r_3} - 1\right) \tag{A3.84}$$

$$a_{86} = \frac{-1}{r_3 C_4} \tag{A3.85}$$

$$a_{88} = \frac{1}{r_3 C_4} \tag{A3.86}$$

$$a_{89} = \frac{-R'}{r_{c5}r_3 C_4} \tag{A3.87}$$

$$a_{91} = \frac{R' r_{43}}{r_{c5}r_3 C_5} \tag{A3.88}$$

$$a_{94} = \frac{R'}{r_{c5} C_5} \left(\frac{r_4'}{r_3} - 1 \right) \tag{A3.89}$$

$$a_{96} = \frac{-R'}{r_{c5}r_3 C_5} \tag{A3.90}$$

$$a_{98} = \frac{-R'}{r_{c5}r_3 C_5} \tag{A3.91}$$

$$a_{99} = \frac{-R'}{r_{c5} C_5} \left(\frac{R'}{r_{c5}r_3} + \frac{1}{R} \right) \tag{A3.92}$$

$$b_{11} = \frac{-1}{2L_1} \tag{A3.93}$$

$$b_{14} = \frac{1}{2L_1} \left(1 - \frac{r_{43}}{r_3} \right) \tag{A3.94}$$

$$b_{15} = \frac{1}{2L_1} \frac{r_{43}}{r_3} \tag{A3.95}$$

$$b_{44} = \frac{1}{2L_2} \frac{r_{23}}{r_3} \tag{A3.96}$$

$$b_{45} = \frac{-1}{2L_2} \frac{r_{23}}{r_3} \tag{A3.97}$$

$$b_{64} = \frac{-1}{\tau_{32}} \tag{A3.98}$$

$$b_{65} = \frac{1}{\tau_{32}} \tag{A3.99}$$

$$b_{84} = \frac{1}{r_3 C_4} \tag{A3.100}$$

$$b_{85} = \frac{-1}{r_3 C_4} \tag{A3.101}$$

$$b_{94} = \frac{R'}{r_{c5} r_3 C_5} \tag{A3.102}$$

$$b_{95} = \frac{-R'}{r_{c5} r_3 C_5} \tag{A3.103}$$

$$c_1 = \frac{R r_{43}}{r_3} \left(1 - \frac{R'}{r_{c5}} \right) \tag{A3.104}$$

$$c_4 = R \left(1 - \frac{R'}{r_{c5}} \right) \left(\frac{r_4'}{r_3} - 1 \right) \tag{A3.105}$$

$$c_6 = \frac{R}{r_3} \left(\frac{R'}{r_{c5}} - 1 \right) \tag{A3.106}$$

$$c_8 = \frac{R}{r_3} \left(1 - \frac{R'}{r_{c5}} \right) \tag{A3.107}$$

$$c_9 = \frac{R R'}{r_{c5}} \left(\frac{1}{r_3} \left[\frac{R'}{r_{c5}} - 1 \right] + \frac{1}{R} \right) \tag{A3.108}$$

$$D_4 = \frac{R}{r_3} \left(1 - \frac{R'}{r_{c5}} \right) \tag{A3.109}$$

$$D_5 = \frac{R}{r_3} \left(\frac{R'}{r_{c5}} - 1 \right) \tag{A3.110}$$

APPENDIX A4: Specifications of the Induction Motor

Parameter	Specification
Rated power	3 hp (2.2 kW)
Motor rated line–line voltage	230 V (rms)
Per-phase stator current	10 A
Poles employed	4
Parasitic resistance of stator	0.712 Ω
Parasitic resistance of rotor	0.8 Ω
Leakage inductance of stator	3.045 mH
Leakage inductance of rotor	3.045 mH
Moment of inertia	0.021 kg-m²
Mutual inductance	128.3 mH

Parameter	Specification
Speed controller (kp)	214
Speed controller (ki)	319.5
Torque controller (kp)	13.45
Torque controller (ki)	13.45
DC voltage regulation (kp)	0.1
DC voltage regulation (ki)	0.025

8 Intelligent Universal Transformer

L. Ashok Kumar

CONTENTS

LEARNING OUTCOMES

At the end of this chapter, the reader will be able to understand:

- The basics of digital automation systems for the future.
- The fundamentals of intelligent universal transformers (IUTs).
- The design features and topology of IUTs.
- Control and optimization strategies of IUT's

8.1 INTRODUCTION

The demand for power is on the rise, and reliable distribution systems are needed to serve this demand. Intelligent monitoring of distribution systems offers a cost reducing approach for providing more reliable systems and for the possibility of catching faults prior to failure. At this juncture, the solid-state state-of-the-art intelligent universal transformer (IUT) will revolutionize distribution system operations in the coming decades. The IUT is a multifunctional programmable power processing device ideal for smart grid and sustainable solutions. The IUT will support multiple functions, such as automatic load monitoring/control/shedding, voltage transformation/regulation, intelligent protection controller systems, power factor control, and nonstandard customer voltages. Furthermore, it will be interoperable with system

DOI: 10.1201/9781003203810-8

controls, will be both smaller and lighter than today's conventional, single-function transformers with the advantage of modularity, and will contain no hazardous liquid dielectrics. As such, IUTs will be the cornerstone of advanced distribution automation and provide the capability for increased customer service options.

IUT will replace conventional distribution transformers with a multifunctional power-electronic system that not only performs the traditional voltage-step function of conventional transformers but also provides numerous other customer services, system operation, and utility control capabilities. In this design, a new protection and monitoring system for dry-type transformers is introduced, which consists of three parts and is based on innovative sensor technologies. The first part protects the device against overheating using an optical fiber sensor, which signals an alarm when local overheating develops. A second unit allows the identification of inter-turn short circuits in the coils by determining and analyzing the transfer function. The third part, which operates independently from the two others, monitors the partial discharge activity and provides the capability for a localization of partial discharge sources. The developed monitoring system operates online, is efficient and cost-effective, and can be easily installed with almost all categories of dry-type transformers. IUT includes overload tracking with the temperature based on surroundings, automated load shedding based on temperature, and/or current levels and predictive overload early warning combined with loss of life estimation. In addition, the tap changer can be blocked if current exceeds the user-defined setting and can prevent load restoration if the hot spot temperature is greater than a user-defined level. There are, however, large savings to be realized by increasing transformer load in a controlled low-risk manner, thus improving utilization of these transformer assets. The ability to protect, monitor, and control the utility transformer assets in one integrated platform is now possible with improved processing power and simplified Windows-driven interfaces. Transformer overload issues are becoming more critical as market demands change and utility systems are pushed harder. In the end, utilities are able to provide a secure reliable protection system that also incorporates overload monitoring and control. This ensures minimal impact on the end user and manages transformer loading risk on the transformer with the potential for introducing economic benefits by improving transformer utilization. In short, it is a revolutionary device that will open new frontiers in the power delivery business.

IUT is proposed as the state-of-the-art electrical key point of advanced distribution automation (ADA) technology. It comprises power electronic construction and a high-frequency transformer. ADA will be tomorrow's distribution system, able to perform the exchanging platform for both data and information, which will be fully monitored and automatically controlled. IUT is introduced as intelligent electronic devices in lieu of traditional transformers. Its controllable architecture enhances reliability and system performance and leads to various services and user benefits like desirable output voltage, level, and frequency, DC voltage options, and sag correction.

ADA, with its new methodology in control and management, leads a gigantic revolution in distribution automation systems, resulting in fully automatic monitoring and control. ADA offers flexible distribution automation, evolving real-time operation and control, and a new approach for exchanging electrical energy, data, and information in a dynamic manner among consumers and system equipment.

Redeveloped electrical construction and open communication architecture empower each other to enable the ADA topology's many benefits:

- Enhancement of the reliability of automation equipment
- Development of intelligent monitoring systems
- Improved efficiency and optimization of system performance
- Optimization of energy management systems in a wide area for system safety, system demand, and power flows

ADA is the cornerstone of convenience integration in moderated electrical technologies from the viewpoint of electrical system designers.

8.2 OVERVIEW OF THE STATE-OF-THE-ART SOLID-STATE TRANSFORMERS

To convert between two different voltage levels, it is necessary to have transformer isolation to fully utilize silicon switches. Figure 8.1 shows the use of a front-end boost converter for harmonic elimination, a full-bridge inverter for conversion of DC to high-frequency AC, a high-frequency AC transformer for voltage-level conversion, a diode bridge rectifier to convert high-frequency AC to DC, and a full-bridge inverter to obtain low-frequency AC. Experimental waveforms are shown at the top of each stage. Consider Figure 8.1 as a building block. The input of the building block is low-frequency high-voltage AC, and the output is a low-voltage AC with the same frequency. The power conversion is equivalent to a step-down transformer except that the power flow is unidirectional. This building block has a solid-state transformer module, and the entire system is connected to the inputs of several modules in series and the outputs of these modules in parallel. The circuit block diagram is shown in Figure 8.2, where the top portion shows three modules using the previously described circuit diagram, and the bottom portion shows how they are connected at the input and output. The main purpose of this configuration is to allow a high-voltage input that can be tied to the distribution voltage level and to have a low-voltage output that can be used for commercial and residential service.

The major difficulty of implementing the circuit shown in Figure 8.2 is to maintain the input voltage balance among different modules. With device mismatching and without any active control, the input voltages among different modules are unlikely to be maintained at the same level. One may argue that adding a set of voltage balancing Zener diodes or clamping circuits will serve the purpose. However, the passive voltage balancing element or clamping circuit consumes a large amount of power

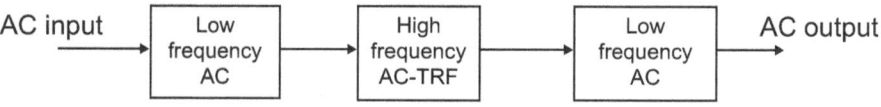

FIGURE 8.1 Solid-state power conversion using high-frequency AC transformer isolation with dual AC outputs.

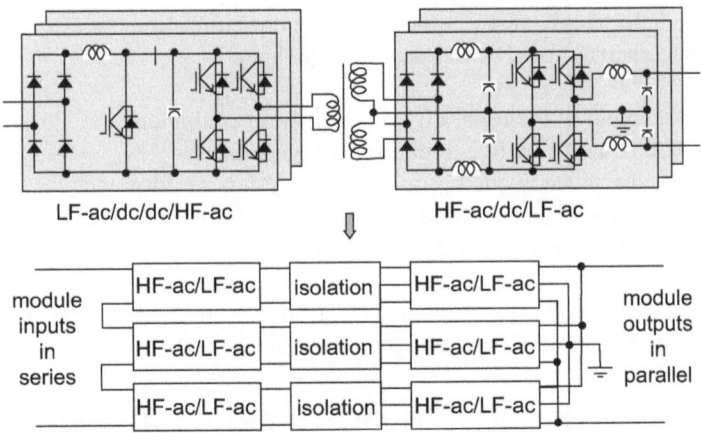

FIGURE 8.2   Modularized solid-state transformer with series-connected input and parallel-connected output.

and is not practical in high-power applications. Since the circuit topology of the basic building block is well-known and is of less concern, the main focus of the entire circuit design should be how to balance the input voltages among different modules. Unfortunately, this crucial problem has not been discussed in the literature. For distribution voltage levels, the number of series-connected modules is typically more than 10. The grounding and insulation between each module require special attention. The common-mode voltage for each module can also create nuisance tripping or faulty operation. Another problem with this circuit is the component counts and their associated reliability issues. With multiple-stage power conversion, each module already requires a tremendous number of devices and components, including semiconductor switches, gate drive circuitry, sensor and conditioning circuitry, and passive components. To accommodate the distribution voltage level with 7.2 kV line-to-neutral or 12.5 kV line-to-line using the previously mentioned series connection for the input stage, at least 15 modules for each phase and 45 modules for three phases are required, assuming that 1200 V semiconductor devices are used for each input stage voltage with 480 V rms. Not only is the balance of each input stage voltage a problem, but also the component count can prevent the system from being practically used.

8.3 ADVANCED DISTRIBUTION AUTOMATION FOR TOMORROW

Advanced distribution automation is distinct from the current distribution automation (DA), which concerns the automated control on basic distribution functions, whereas ADA is proposed with fully automated control capability with regard to each of pieces of controllable equipment in the distribution system. This provides the system performance improvements by adding functionality and enhancing reliability, quality, level of security, and availability in contrast with the today's distribution operations.

ADA brings a revolutionary modification in infrastructure of traditional distribution systems, which will occur in a manner similar to the rate of investment in heritage systems and the rate in which technology will progress. IUT can serve three functionalities: first, as a smart transformer for distribution substation; second, as an intelligent device with the desired outage related to commercial customer requests; and third, for overhead lines in distribution system. Two critical points in the concept of ADA are that the flexible electrical architecture and the open communication architecture improve advance monitoring and control functionality for an interoperable network of components. These two parts synergistically empower each other to comprise the forthcoming distribution system. ADA is introduced as the heart of a smart power system especially in the case of the delivering procedure. Flexible electrical architecture employs intelligent electronic devices (IEDs), distributed resources (DR), and new electrical and electronic technologies. On the other hand, communication and control systems that are based on open communication architecture provide the suitable platform for information exchange. Both of these, together with the real-time state estimation, provide the most applicable tools for predictive simulations and real-time optimization performances like demand and energy management, reliability, efficiency, and power quality control. ADA employs novel moderated power electronic technologies to stimulate the creation of new advanced technologies. We discuss one the most important ones that takes the critical key point in ADA as the intelligent universal transformer (IUT).

8.4 INTELLIGENT ELECTRONIC DEVICES

The intelligent electronic device (IED) is an important key point in contributing to ADA in the near future. IED is defined as an intelligent component that is incorporated among one or more processors for receiving or sending data or controlling command from or to external sources. In this regard, the multifunctional meters, digital relays, and IUT are the well-known IEDs introduced for the next generation of distribution automation. Successful integration of the IEDs should be integrated with the overall distribution system, which involves protection, control, and data acquisition. This leads to a reduction in capital and operating costs and brings out panel reduction. It also releases designers for providing redundancy in equipment and databases.

8.5 INTELLIGENT UNIVERSAL TRANSFORMER

The most strategic equipment in ADA is the intelligent universal transformer (IUT), proposed as an IED in the near future.

8.5.1 IUT Basic Construction

Recent progress in silicon carbide (SIC) materials and their applications has led to the development of HV-HF power electronic devices with a 10kV, 15 KHZ switching pattern that extends the PWM usage in high-voltage applications. Also expected is the revolutionary utilization of power electronic devices in many applications, such as motor drive, military applications, and power distribution and conversion. The *Electric Power Research Institute* (EPRI) effort for IUT programs opens a new way to transfer power in a controlled manner, especially in the case of distribution systems.

The application of power electronic transformers for power distribution system, which is used in IUT construction for tomorrow's distribution system, has been discussed in recent years. Actually, the IUT is a state-of-the-art power electronic-based transformer that not only steps voltage in lieu of current conventional distribution transformers but also acts as an intelligent controllable device with the major advantages of DC output or multiple frequency options, the ability for converting from single-phase to three-phase, and improvement of power quality functionalities like voltage sag correction, harmonic filtering and oil elimination, reduction in weight and physical dimension, and interoperability. These all enable the IUT to serve as an intelligent multifunctional node for the upcoming distribution automation. IUT design is based on a high-frequency transformer (typically a few kilohertz) with isolation functionality. This construction leads to core size reduction that is impossible with traditional 50 Hz transformers.

In general, the IUT will be based on a high-voltage/low-current power electronic base transformer instead of traditional distribution transformers. At first, in the primary side converter, the input low frequency (50 Hz) sine wave voltage is converted to a square-wave high-frequency AC that will be magnetically coupled to the secondary stage as the isolated high-frequency voltage. Then at the secondary side, power converters will change the waveform from HF voltage to LF (50 Hz). This occurs by the synchronous operation of the primary and secondary side converters by a 50% duty ratio modulating in the switches of HF square wave.

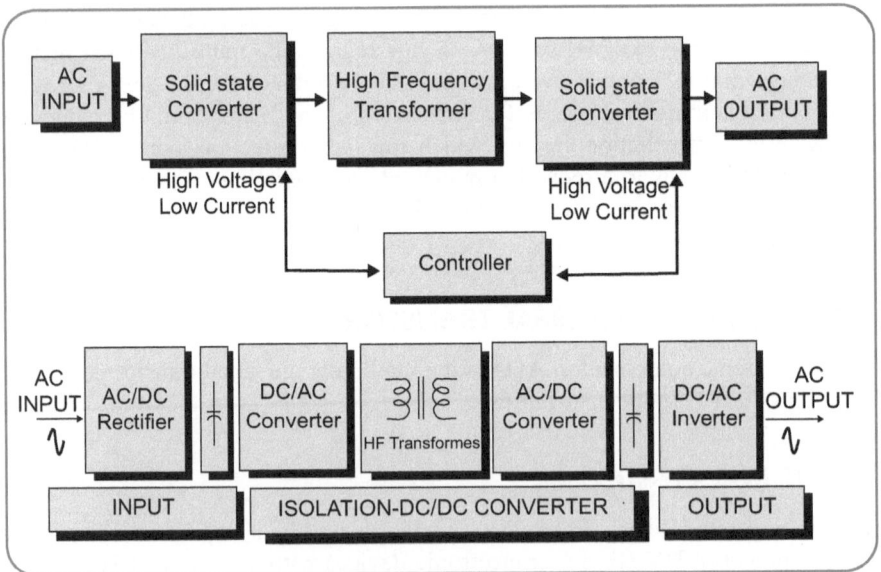

FIGURE 8.3   IUT basic component diagram.

The IUT communication interface is based on the IEC 61850 as the open communication standard for exchanging information and data that will be integrated by EPRI in the IEC-TC-57 for standardization of the object model. IUT offers numerous benefits and advantages, such as delivering new service options like automatic sag correction, DC voltage service, reliable diverse power as 400 Hz service for use in communication, accessibility of three-phase power even from a single-phase source line, availability for storing electrical energy, the capability of voltage regulation in real-time operation, harmonic filtering, flicker mitigation, and dynamic system monitoring.

8.6 IUT OVERVIEW

In order to overcome the voltage sharing problem for series-connected devices or converter stacks, several advanced multilevel converter circuit topologies have been discussed. By incorporating advanced multilevel converters with the use of HV-Si devices (HV-IGBT were used for the bench model, and HV-S-GTO is used for the field prototype) and SiC devices in the near future, it is possible to develop power converters for distribution system applications. For the field prototype, EPRI proposes the concept that combines a novel current mode resonant conversion circuit (rather than hard switching) with *super gate turn-off thyristor* (SGTO) devices (rather than IGBTs [*insulated gate bipolar transistor*] or power MOSFETs) and SiC-based Schottky diodes. Together, these provide low-loss, highly reliable, low-stress circuits, resulting in smaller and lighter equipment.

The proposed design employs a circuit concept where multiple cascaded modulated H-bridges convert the 13.8 kV, 50Hz AC into 3.5 kV DC at near unity power factor. Each 3.5 kV DC is then inverted into single-phase 20 kHz AC by a series resonant converter using SGTO switches and Si-C diodes. Voltage level transformation is achieved through a high-frequency transformer. The resulting high-frequency AC output of the transformer is rectified to 400V DC using a diode H-bridge. The 400V DC is then inverted to center-tap 110/230V single-phase 50 Hz AC using a hard switched inverter bridge. The outputs of the diode H-bridges are connected in parallel to achieve the desired power level. Each resonant converter output is a current source to the load, so these are also connected in parallel. The high-frequency transformer also provides galvanic isolation. This configuration of resonant converter that uses the leakage inductance of the high-frequency isolating transformer as the resonant circuit L is inherently suitable for high-voltage isolation applications since the required leakage inductance makes it possible to use a substantial isolating layer between input and output windings. The IUT design also employs external natural air cooling, eliminating the expenses and problems associated with liquid cooling. SGTO can further improve power density and cost by allowing the switching frequency to run up to 50 kHz (compared to IGBT limitation of 20 kHz max). SGTOs also have lower conduction losses (50% that of IGBTs) and lower switching losses (50% that of IGBTs). In addition to lower losses, SGTO devices have better thermal management (lower thermal resistance) compared to IGBT or other power switching devices, ultimately making the equipment smaller, lighter, and less expensive. Since IGBTs do not have wire bonding in their package, SGTO modules have 100 times better reliability than IGBT modules. SPCO's concept further optimizes the design by using Si-C antiparallel diodes with almost no switching losses. The design concept is modular and scalable, providing quality, reliability, and economy of scale.

8.7 IUT DESIGN FEATURES

The IUT follows a modular design that incorporates the following features for performance, ease of operation, installation, troubleshooting, and life cycle cost.

- Use of low-loss and reliable SGTO devices
- Higher frequency switching for a compact footprint
- Resonant switching for less stress on devices
- Modular design
- No mineral oil
- Sealed compartment for electronic components
- Robust construction for outdoor applications

The power electronic components are arranged in four separate stacks, each rated at, e.g., 25 KVA. Each stack consists of an active front end (a level of the cascade converter) and a resonant converter. Each stack has a high-frequency transformer. One stack includes the output inverter. The control boards and bus bar then overlay the power electronics. Each 25 KVA stack contains all necessary electronic components such as capacitors, inductors, and resistors that are required for operation of the system.

8.8 IUT TOPOLOGY

IUT topology is comprised of power-electronic-based devices that act as rectifiers and inverters, together with a high-frequency (HF) transformer. The HF transformer ratio is 1:1, and it acts as an isolation device. The control unit in Figure 8.4 represents the intelligent transformer with controllability in the input–output stages for providing the desired output level options and employing optimization tools for eliminating the harmonic disturbances in output and input.

This topology is comprised of seven individual blocks. The multilevel rectifier (1r) and multilevel inverter (1i) in first stage rectify and convert the AC input voltage to high-frequency square wave. The DC bus and capacitors are installed in the second stage. A high-frequency transformer in the third stage with a 1:1 ratio has the duty of isolation. At the fourth part, DC output rectifiers and filters make up the full bridge inverter, involving the split DC bus capacitors. The main inverter for delivering outputs with the specification of 110/230V AC, 50 Hz is in the fifth stage, comprising the full bridge inverter and single DC bus capacitors. The auxiliary inverter is in the sixth stage, providing 400 Hz output consisting of full bridge inverter and single DC bus capacitor. Finally, the DC–DC converter is considered for 48V DC output. The controllable parts are the rectifier, HF inverter, and secondary output inverters represent as an intelligent unit. Advanced control methodologies are introduced for IUT

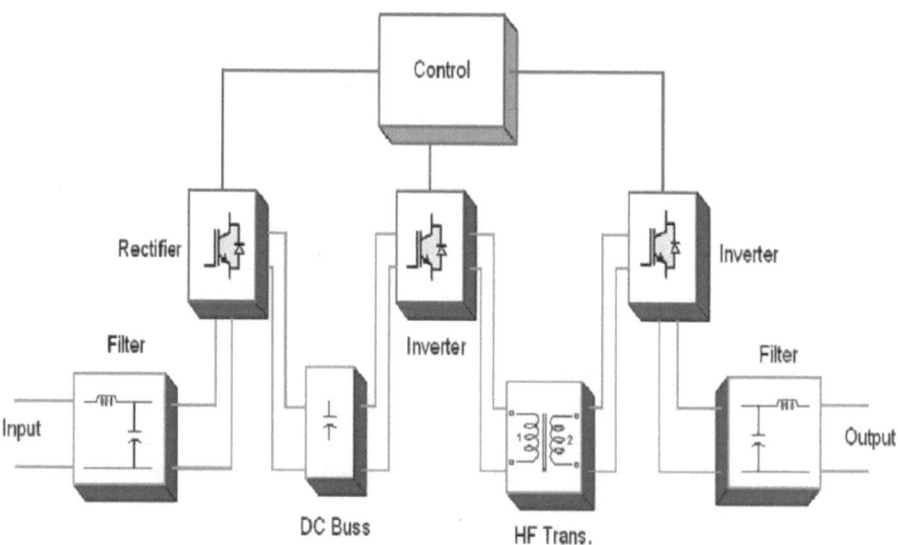

FIGURE 8.4 IUT power electronic construction.

controllers. The neural controller is fully described in proposing the online adaptive
control via an artificial neural network. A developed adaptive neuro-fuzzy inference
system is also considered in controlling IUT. In this approach, the gate array (GA)
optimizes the firing angle for more smooth and robust control. The HF inverter in the
input stage and output inverters are introduced for the GA control strategy. Several
topologies have been described for IUT. We choose the four-layer topology based
on seven basic blocks involving the multilevel rectifier and inverter, high-frequency
transformer, and the four-layer outputs comprised of DC voltage, AC 400 HZ, and
two main 230 V AC, 50 HZ.

In the first stage, the multilevel rectifier (1r) and multilevel inverter (1i) rectify the
input AC voltage and convert it to HF HV square wave. DC voltage is produced in
the second stage by means of the DC bus capacitors. In the third stage, the HF transformer takes on the task of isolating inputs from outputs. Fourth, rectifiers and filters
make the DC output voltages. In the fifth stage, a main inverter assigns 110/230 V,
50 Hz output. The 400 Hz output service is developed by an auxiliary inverter, and a
DC–DC converter takes the 48 V DC output. In this topology, IUT defines the driv-
ers' outputs services.

As it is clear that IUT can be developed to deliver DC output voltages at any
desired level, AC voltages with different arbitrary frequencies were impossible with
traditional transformers. On the other hand, in IUT, the transfer ratio is at unity,
so the weight and physical dimension in the IUT make a sizable reduction possible because it can be considered in connection with oil-free technologies, with
the results of its being maintenance free able to lower pollution and contribute to a
cleaner environment. The entire transformer is considered an energy transformation

device, so that the instantaneous power across the input terminals is equal to the output terminals.

8.9 FLC MODERN CONTROLLERS

Fuzzy logic controllers (FLCs) are nonlinear controllers with a widespread application on unknown, linear and nonlinear, simple and complex systems. It controls systems without any information regarding the transfer function among the input and output variables. It acts on human-based rules in sentences for producing the control strategy based on rule equations that come from human experience. The fuzzy logic controllers can be progressed amid noisy, imprecise input, so they are really more effective and easier to implement.

In a FLC, the input variables are mapped by sets of membership functions known as fuzzy sets. A fuzzy set comprises a membership function that can be defined by parameters. The value between 0 and 1 reveals a degree of membership in the fuzzy set. The process in which the crisp input values convert to a fuzzy value is termed fuzzification. The fuzzy operations and rule-based "inferences" collaborate to describe a so-called fuzzy expert system. Traditional control systems are structured on mathematical models that employ differential equations comprising the system reaction to the inputs. Fuzzy logic controllers operate from an input stage, a processing stage, and an output stage. In the input stage, based on inputs from sensors, switches are mapped to the proper membership functions. In the processing stage, each appropriate rule is invoked, and a result is generated for each of them. Finally, the results are combined to form the rules. In the output stage, the combined result is assigned a special control value.

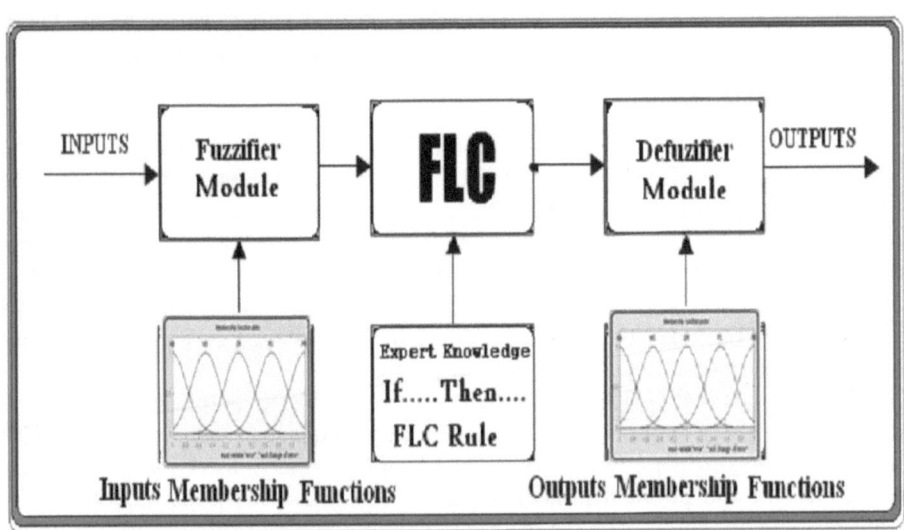

FIGURE 8.5   FLC basic module including fuzzifier, defuzzifier, and fuzzy sets.

8.10 CONTROL AND OPTIMIZATION STRATEGY

The control strategy is based on controlling the angle of the inverter in the input–output stage of IUT using GA optimization for harmonic eliminations. In the input stage, the IUT is directly connected to grid. For eliminating harmonic distortion, input current is sensed and controlled by the GA to maintain the input current being in phase with input voltage. This action yields to unwanted harmonic elimination. Each of the inverters in the output stage is controlled and optimized trough the GA, which employs three basic rules—selection, crossover, and mutation—for reproducing the next generation based on current population. At the first stage, the individuals representing parents are selected to contribute the next generation of population. In the second stage, the crossover rules combine two parents contributing children in the next generation. Random changes in the individual parents for contributing children are performed in mutation rules. In accordance with IUT topology, the variables of control scheme corresponding to the chromosome are the controllable switching angles defined as firing angles for the input stage and each of the four outputs of IUT.

Control strategy is based on fuzzy logic controllers in the input and output stages. In the primary stage, the IUT is connected directly to the smart grid. For eliminating the harmonic distortion, the input current and voltage should be both sinusoidal and in phase with each other; this can be achieved by fuzzy logic controllers. In the input stage, the FLC in AC/DC converters senses the input current in IUT and compares it from the reference current, and any deviation from the reference is mentioned as error. Error and change of error are considered inputs of the FLC, which assumes the duty of control for preventing any disturbances of input from the grid.

8.11 CONCLUSION

The rapidly rising costs of conventional transformers, the need for reducing the size and weight of the magnetic components, the need for distribution automation and monitoring to improve reliability, the requirements for new services, and the need to meet customers' power quality and reliability requirements have driven possible alternative solutions to the conventional transformer. The development of the IUT provides an opportunity to offer DC service and high-frequency AC service, along with the ability to convert single-phase service to three-phase for supplying customer loads in areas not served by three-phase circuitry. The ability to provide a wide range of services and improved operational benefits has put solid-state transformer technology—an intelligent universal transformer (IUT). The IUT represents a promising new technology that can impact a wide range of applications, and the FLC control methodology is concerned with overcoming ambiguous conditions, nonlinear and complex systems, and with enhancing robustness for the new modern technology described as IUT. DC and three-phase output voltages are the benefits of using four-layer IUT topology. In this, the four FLC controllers take the role of control, guarantee stability, and guard the whole system against disturbances in the input–output stages. It also leads to efficiency enhancement in system performances. ADA infrastructure has been raised in connection with the future needs of the next phase of distribution automation. It is directed toward full network functionality. Reliability enhancement is a part of this innovation and can be described using modern adaptive solutions for forthcoming projects, especially for IUT in the smart grid of future.

9 Role of Industry 4.0 in Energy Management Systems

J. Jayakumar

CONTENTS

LEARNING OUTCOMES

At the end of this chapter, the reader will be able to understand:

- The basics of the energy management system, Industry 4.0, and data analysis.
- Concepts of Industry 4.0 in energy and utilities management.
- The need for advancements in energy management systems.
- The roles of Industry 4.0 in energy management systems.

9.1 INTRODUCTION

Nowadays, industrial manufacturing companies are expecting and facing strong demand to increase their productivity by realizing smart factories and smart manufacturing. Machine automation, monitoring of equipment and machines, predictive maintenance, production traceability, and energy management are essential and inevitable. Given that human efforts lead to error and consume much time, this chapter presents ideas about the importance of automation in energy management with the help of Industries 4.0 ecosystem. There were Industrial 1.0, 2.0, and 3.0 versions, and these numbers generally indicate the number of revolutions that have occurred in the industrialized period. So the first ever industrial revolution, or Industrial 1.0, occurred in the 18th century when mechanical production was invented, with steam- and water-turbine-powered equipment back in the United Kingdom. The second industrial revolution, Industrial 2.0, began when Henry Ford invented what we

DOI: 10.1201/9781003203810-9

all know now as the production assembly line or mass production assembly line, wherein intensive labor is involved, and electrical energy was used to automate, i.e., control, systems. In Industrial 3.0, which occurred around the 1970s to early 2000s, we saw a great deal of automation of particular production processes in the factory. This is the direct result of inventions like the internet and Wi-Fi, and we could now more or less automate a few processes in the factory. Finally, what we have today is Industry 4.0, which revolves around technologies like intelligence production incorporated with IoT, cloud technology, and Big Data. Now, the Industry 3.0 and Industry 4.0 standards may seem very similar, so in our introduction, we look at what is different between them. Now, Industrial 3.0 is a centralized system.

The control of each machine is very much centralized. An IT system automates individual machines and the process. There is no communication between two machines, the connectivity is more or less limited inside the factory, and dedicated lines are required to run data between each machine and the control room. In Industrial 4.0, the centralized control system does not exist the system is very much decentralized. Things like IoT and cloud technology come into play, and they start automating not just one task but multiple tasks across the entire production line. So we have machines talking to other machines depending on conditions, enabling the gathering of more data and making the entire process more intelligent. This enables us to do something called flexible or just-in-time manufacturing, wherein the demand and the supply are perfectly matched. The additional benefits include more sensors and information access, more data for better decision making, and remote access from the factory machines. We will look at other benefits in the next section. Now, the major benefits include productivity and output; i.e., when you are manufacturing exactly to the demand and you have contact directly to the customer, you can improve your productivity by reducing downtime and reducing switchover time. The second point is enhanced customization for flexible manufacturing; i.e., the machines can be adapted very quickly depending on what data are gathered.

In most cases of Industrial 4.0, the customer directly talks to the factory; an intermediary is never involved, and because of that, the machines know what products to make immediately when they received the order. The second point to be noted is that, because so many sensors are involved, the machine can predictively maintain itself in the sense that whenever a fault is about to occur, the machine can warn factory workers, and preventive maintenance can be made. Downtimes and fatalities can thus be reduced. Improved worker safety is another point, which is a direct result of the previous two points. Because of most of these processes are automated, a smaller number of people are working on the factory floor, which directly results in fewer mishaps and fewer injuries. One final point is that is access to data across the chain. As previously mentioned, most of these Industrial 4.0 limitations have customers directly talking to the machines, and this does not stop there. Even the delivery people can know exactly what is going on in the factory, so that they can come to the factory right on time to take those orders and deliver them in a timely fashion.

Now, the major technology that enables most of these things is something called the IIoT, which stands for Industrial Internet of Things. Most of us must have heard about IoT, which stands for Internet of Things. The Internet of Things connects a few smart devices in our house, which predictively collect data and make our dietary decisions a little easier. A good example is a smart home system like Google Nest, which connects to your thermostat and to your smart phone, a few lights, and fans. I then maintains a good temperature inside your house, turn the fan on and off whether you are there are not, and would let you operate certain smart home systems when you are away at work.

A good example is a smart water heater, which you could turn on when you are at work; by the time you reach home, the water would be hot enough. This is IoT, which is very consumer specific. It is targeted on the end user, whereas IIoT, as we will see, is targeted on industries, as the name suggests. Moving on to the Industrial Internet of Things, the general usages are completely different from those of IoT. Whereas IoT is more of a consumer-oriented product, IIoT is an industrial process that involves manufacturing, supple chain monitoring, and management systems. Now IIoT uses more sensitive and more precise sensors, including more locations of the technologies. In other words, the sensors that are used in the IIoT application are more customized and more robust, so that we can have reliable information.

What are the similarities between IoT and IIoT? IoT is for household usage in the commercial sectors, and IIoT is for an industrial sector. They both have data management and connectivity and data security, and they both require a secure cloud—but that is basically the end of their similarities. After that, the two are very dissimilar.

Finally let's look at an example of Industrial 4.0, specifically at the smart parking example. Smart parking is basically something that we knew existed, but it was truly revolutionized after the invention of Industrial 4.0 and 5G technology. The basics of the smart parking system is that a sensor detects whether the car is parked or not, and this data can be transferred to a centralized control system or a server. Now, this server can be placed anywhere outside the city or inside the city, and the server communicates with this server via 5G or a Wi-Fi system. A smart phone app lets users know whether a certain parking space is free or not.

For example, let's say we are going into the mall at 3 o'clock in the afternoon. An app would tell you whether you would find parking spot or not at the right time, and you could accordingly adjust your traveling time and plan accordingly. The parking fee can be paid directly through the app or when you park on the spot. The app notifies you about the parking fee, and when you finally go back to your car, the app finishes your payment automatically for you and lets you how much to pay, and because of the various implementation technologies, you can even ID cars. For example, you can tell one car from another. You can even have cars identified on the basis of who owns it. So, for example, in a residential area, you do not have to ask people to pay for parking if they already live in the area; residents' cars can be tagged, so that parking fees are not levied for people who live in the area and for other vehicles. For example, when a special goods vehicle requires more space in maintenance, it can be charged a little more than usual. With this system, you can detect whether someone is parking in a no-parking zone or if someone is parking during off-hours, and such data can be transmitted directly to the police or the traffic control system, thus basically ensuring the credibility of the system. This an excellent example of an Industrial 4.0 implementation, where a sensor is location aware, is making decisions by itself, and is not centralized. Each sensor is connected individually to the network, so even if one sensor dies, that doesn't mean that the entire system goes down, or if one mobile tower collapses, that doesn't mean the entire system collapses—there is always another mobile tower. This is an example of Industrial 4.0.

9.2 CONCEPTS OF INDUSTRY 4.0 IN ENERGY AND UTILITIES MANAGEMENT

There are four important concepts of Industry 4.0 in energy and utility management application.

9.2.1 EXTENSIVE MONITORING

The development of technologies for instrumentation and monitoring of industrial processes enable data capture in ever increasing resolutions, allowing increasingly powerful analyses. In energy and utilities management, sophisticated physical meters (instruments) are capable of interpreting physical quantities that allow for the understanding of processes of interest, monitoring variables that range from applied power, for example, to harmonics that describe the quality of the electricity consumed.

In addition to technological advances, the costs of acquisition and installation of modern sensors and instruments have become increasingly accessible, allowing broad and deep understanding of the characteristics of industrial processes of interest, allowing redundancy of measurements and the obtaining of high-quality data—essential for planning, control, and improvement of energy efficiency and operational efficiency.

9.2.2 INDUSTRIAL INTERNET OF THINGS (IIoT)

The Internet of Things is another widely discussed concept and refers to an entire "network of physical devices that include sensors, actuators, electronics, and connectivity, allowing the integration of the physical world with computer systems." In our context, the Industrial Internet of Things, a term often used as a synonym for Industry 4.0, refers to the application of technologies such as Machine Learning and Big Data to exploit sensor data, communication between machines (M2M), and automation systems to improve industrial and manufacturing processes.

In energy and utilities management, Industry 4.0 is realized in the connectivity between measuring instruments and the entire information and automation architecture of industrial organizations, extending the capacities for collection, communication, and storage of large volumes of data related to the consumption, generation, and transformation of energy input.

9.2.3 ANALYSIS OF LARGE VOLUMES OF DATA

Typical industrial applications can involve thousands of meters collecting data at high frequencies and generating gigabytes of data each day; in energy quality applications, for example, specialized meters today visualize the network each millisecond.

This abundance of data and the increasing availability of computational resources allow the application of specific techniques of artificial intelligence with the aim of facilitating the prediction of variables and the identification of patterns of interest in a range of industrial processes.

Due to the very nature of the phenomena that produce data collected from industrial operations and the limitations of the instruments used to capture them, the development of prediction models based on data collected from industrial operations involves considerable levels of noise and imposes additional pressures on the volume, variety, speed, and veracity requirements of the data, something common to Big Data applications. Efficient algorithms for processing data quality are thus becoming as essential as algorithms for the construction of prediction models.

In energy and utilities management, the data available can give rise to, for example, prediction models for energy consumption (or energy generation) of operations, starting from planned production levels or other contextual variables; models for learning and establishing the ideal modes of operation, which permit effective levels of energy consumption; models for analyzing the energy efficiency of processes, starting from the capture of entry and exit variables and knowledge of the transformation phenomena involved.

9.2.4 EFFICIENCY AND SUSTAINABILITY

Behind the entire investment in Industry 4.0 lies a common objective: increasing the efficiency and competitiveness of an operation. The benefits are direct and carry the potential to establish a virtuous cycle of investment, result, and reinvestment: more competitiveness results in better financial results; with more cash in hand, more investments can be directed to capacity expansion, productivity technologies, operational efficiency, and energy efficiency; greater efficiency ensures lower levels of greenhouse gas emissions, reducing environmental impact in addition to improving the quality of work, both of which positively impact the community.

9.3 INDUSTRY 4.0 AND ENERGY AND UTILITIES MANAGEMENT

Energy management is one of the main pillars of Industry 4.0. The motivation comes from a combination of environmental aspects, cost pressure, and regulation, as well as the proactiveness of organizations when it comes to efficient consumption of energy and utilities. In addition, the integration of different sources of energy generation in an increasingly demanding and distributed market requires management technologies capable of recognizing, predicting, and acting in a way to guarantee quality, sustainability, and efficiency, including costs in energy consumption.

Modern energy and utilities management systems should be able to exploit a large volume of data collected by various types of meters on a number of variables of interest for a certain industrial operation, assembling the preceding concepts—extensive monitoring, the Industrial Internet of Things, analyses of large volumes of data, and efficiency and sustainability—around a common, integrated, and robust objective.

10 Nonintrusive Load Monitoring for Efficient Energy Management and Energy Savings

Krishna Rubigha and Dhiksha Mohan

CONTENTS

LEARNING OUTCOMES

At the end of this chapter, the reader will be able to understand:

- Nonintrusive load monitoring (NILM) and its functionalities.
- The components and working principles of NILM.
- The advantages and challenges of NILM.
- The future scope for NILM.

10.1 INTRODUCTION

With amplifying energy consumption around the world, profound and substantial efforts are being taken for better and efficient energy management systems [1]–[6].

DOI: 10.1201/9781003203810-10

Nonintrusive load monitoring (NILM) allows the user to measure energy consumption and monitor the individual loads of a system (residential/commercial/industrial) without multiple sensors [4]. This cuts the cost involved in the installation of sensors [5] and data acquisition devices for individual loads, thereby increasing the efficiency of energy management practices.

This approach, also known as nonintrusive appliance load monitoring (NIALM), is currently being widely used for residential and commercial energy monitoring and in smart buildings [4], [6]. It helps in monitoring the entire building's energy consumptions through a single device. Hence a device is used to measure the total electrical energy consumption from the main electrical panel, and the aggregated data are broken down to find the consumption of each appliance using advanced AI (artificial intelligence) algorithms. This allows the user to understand the accurate energy consumption of each appliance with high precision [7]. Thus the user can obtain home insights, energy reports, and measures to be taken to increase energy savings. This chapter discusses the methodology, hardware requirements, data acquisition and processing, challenges, and areas of the future scope of NILM.

10.1.1 How It Works

Each electrical appliance has a unique electrical signature. Hence, we can think of the measurement by energy meters as "the collection of unique electrical signatures.'"

Hence, with a single device, the data of total electricity consumption are sensed and used for further analysis. The electrical current drawn is measured at a high-frequency sampling rate (over 1 million times each second) [8]. This large amount of data is classified using machine learning and data disaggregation algorithms. Energy disaggregation is the science that differentiates energy data into individual appliances.

10.2 DATA ACQUISITION FOR NILM

Data acquisition and data processing play a major role in the NILM system for monitoring load behavior. With the help of smart meters/sensor clamps, the voltage and current consumed is measured. These are the fundamental parameters needed to disaggregate the data, and the latent parameters are obtained using intelligent algorithms. The latent parameters that can be calculated from these are real power, reactive power,

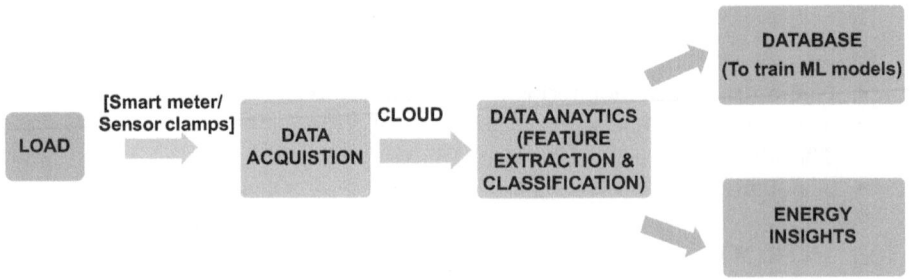

FIGURE 10.1   Flow diagram of NILM system.

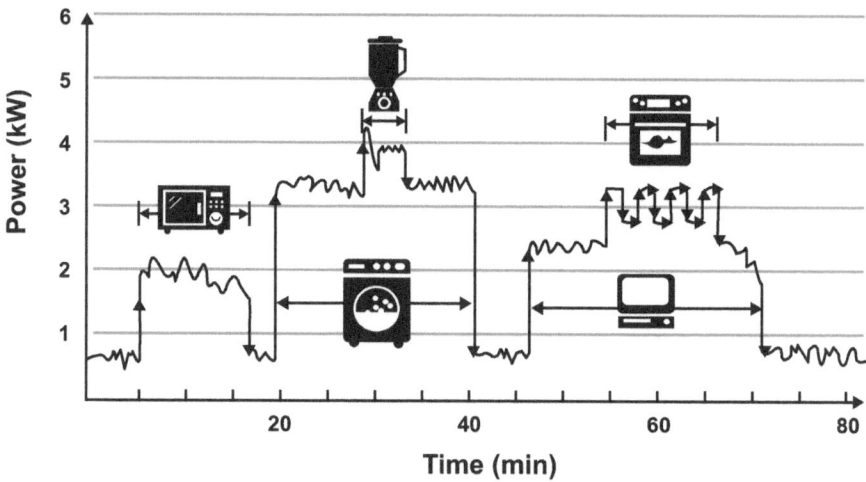

FIGURE 10.2   Graphical representation of electrical load signature classification [4].

power factor (PF), total odd harmonic distortion (TOHD), total even harmonic distortion (TEHD), current harmonics, current wavelet transformation, instantaneous admittance, instantaneous power, current eigenvalue, transient power, and VI trajectory, etc. [3]. The collected data are transferred for processing through wireless communication.

10.3 DISAGGREGATING THE DATA

Energy disaggregation is "the process of breaking down an 'aggregated' energy signal into each device that is contributing to this total amount" [9]. The energy data are disaggregated by machine learning algorithms, Big Data processing, and real-time analytics. Several computational techniques are used to identify electrical signatures of each appliance to produce insights [10].

Different energy companies and researchers use diverse algorithms for disaggregating the data. Hence the sampling rate and amount of data required vary for each application, system, and use case.

10.4 DATA STORAGE

- The acquired data is stored and processed in the cloud server.
- Private cloud servers are currently being used to store the acquired data and the delivered energy insights.

10.5 COMPONENTS

10.5.1 Data Acquisition Devices (Sensors)

The two main parameters, voltage and current, are measured by the sensors. Voltage transformers (voltage clamps) are generally used to sense the voltage. For microscopic feature metering, the critical component to be considered is the current sensor. Sensors like current transformers, Hall effect sensors, GMR sensors, Rogowski coils, etc. [3] are widely used for sampling current signal at a high rate. The sensors have to acquire high-resolution data to monitor the data real time [3].

Based on the sampling frequency rates and the latent parameters to be calculated, the metering hardware platforms are divided into two types [3].

1. **Macroscopic metering hardware:** This is used for sampling data at low frequency. Hence it inexpensive as it involves lower communication and data processing charges.
2. **Microscopic metering hardware:** This is used for sampling data at relatively higher frequency. It provides the accurate detection of load behavior and is expensive.

10.5.2 Hardware Considerations

Mainly there are two ways to monitor the loads nonintrusively, based on acquisition of data:

1. Acquiring data using advanced metering infrastructure (AMI)
2. Acquiring data directly from the main electric board (by clamping the sensors)

Smart Meter

In this type, a smart electricity meter is installed that measures the electrical data and communicates it through wireless transmission. The received measurements are further processed, disaggregated to calculate the latent parameters, which will be discussed later in the chapter. The processed data are displayed to the user in house displays or mobile applications. The displayed data contain graphical representations and report the energy consumed with recommendations to save energy in that system. The data reported are also exchanged with other platforms like e-mail, android applications, online dashboards, etc.

The sampling rate of smart meters is very low (they sample data once in half an hour) when compared to NILM requirements. This issue is solved by the integration of a CAD (consumer access device), which prevents the loss of data during transmission. It enables cloud connectivity for smart meter data by sending it to the designated cloud service. It can be integrated with meter data acquisition software for "advanced smart metering functions" [2]. It supports real-time monitoring of energy data and device management.

High Resolution Sensors

Presently, various patented sensors are in use for the acquisition of voltage and current signals. Conventional electricity sensors require hardwiring and complex installation, whereas the latest patented sensors of various energy companies are compact [11], flexible, and reliable, and they sample data at an incredibly high rate (>10,000 times per second [7]). These require simple installation; one sensor per building is sufficient for its operation.

The sensors are installed in the electric panel through magnetic probes connected to the circuit breaker and the main current table to sample the voltage and current signals.

Wireless Communication and Other Modules

A wireless communication module is another major component for data transmission. Technologies like Zigbee and Bluetooth are widely used for energy data transfer.

An MCU (microcontroller unit) is needed for managing the operation of data collection, transmission, and processing.

Note that different energy companies develop different patented sensors, which may include all the previously mentioned components in a single device, leading to compactness and reliability.

Power Supply

Currently, the advanced metering infrastructures and sensor clamps are powered by the main power supply, and battery power may also serve as an auxiliary power supplement [3]. The transformation of electrical voltage (220 V) is suitable to power them with the help of additional components [3] (rectifiers and voltage regulators).

10.5.3 Software Requirements

The data acquired through sensors/smart meters are broken down by various data analysis methods and machine learning models. The energy data of each appliance are collected beforehand by installing individual sensors. This helps in training the machine learning models to recognize the electrical signature of that particular appliance. Later, these data are compared with the total electricity consumption data

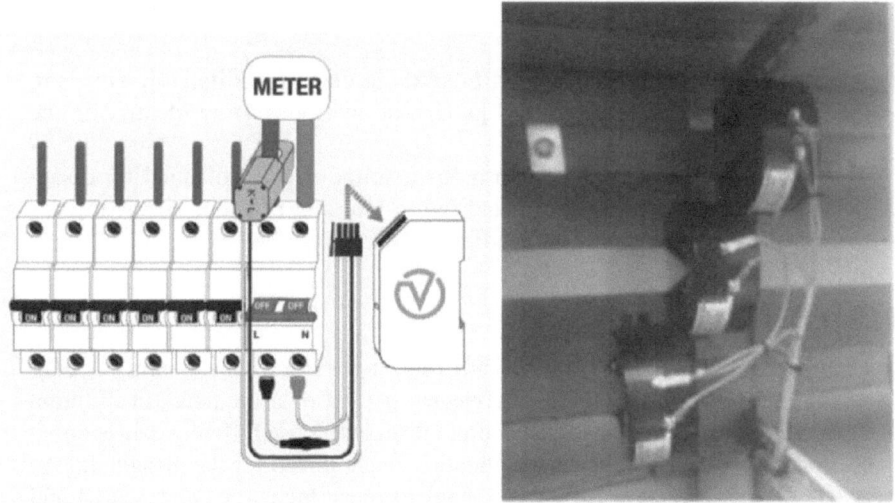

FIGURE 10.3   NILM hardware model [12], sensor clamps for NILM.

obtained from the smart meters/sensors for disaggregation. Appliance fingerprint detection algorithm [7] is one such algorithm currently being used.

Also, an open source NILM toolkit is available to help researchers evaluate the efficiency and accuracy of the developed NILM algorithms [2]. At the interface between hardware and cloud server, a CAD device is required if AMI is used for a NILM application. It enables cloud connectivity for smart meter data.

10.6 ADVANTAGES OF NILM SYSTEM

1. **Effective energy monitoring:** Effective energy monitoring eliminates the cost and installation of multiple sensors to monitor each appliance individually, hence significantly increasing the reliability and efficiency of energy management. Records provided by energy companies also prove that monitoring the loads by NILM increases the energy savings by 15% in the residential sector [13].

   In India, the Smart Meter National Programme is implemented to replace conventional meters with smart meters (approximately 17,45,665 smart meters installed). This facilitates the digitized and accurate measurement of energy consumption. Though the Programme provides automatic and accurate electricity consumption bills [1], it doesn't benefit the consumers in terms of smart monitoring. Adopting NILM systems will be beneficial for better analysis, monitoring, and conservation of energy consumption.

2. **Reduced carbon footprint:** NILM has also found to reduce carbon footprint [7] through better insights by streamlining the energy and increase real-time safety.

3. **Predictive load maintenance:** Through NILM, the appliances can be constantly and remotely monitored to detect any inefficiency [7]/need for maintenance.

4. **Long-term energy monitoring benefits in energy audits [14]:** This is in addition to the well recognized pattern of energy usage in identifying the areas of energy loss.

5. **Aid in advancement:** The enormous amount of data collected for disaggregation can be processed in the future building of better NILM models and ML algorithms, thereby leading to deeper insights

10.7 CHALLENGES

• The major challenge in building NILM systems is the unavailability of public data sets [5] containing the electrical load measurements in different applications. This is because different researchers use different data acquisition devices with different specifications customized for the disaggregation algorithms they use. Hence it is hard to reuse them for our research and applications.

- If there are two similar appliances, it's becoming difficult to distinguish their electrical waveforms. In the case of high-voltage load monitoring, the energy measurement of appliances with very small power consumption becomes almost negligible. This leads to faulty detection of loads during monitoring.
- During execution, the cost of data transmission and storage is very high as the data sets are too large and complex. The cost should not exceed the cost of installing an extra set of current transformers/submetering, which is the present method.
- It's hard to develop a prefect algorithm to map the electrical signals to the target device, as it involves distinct devices, complex electrical signatures, and several parameters [15].
- The industrial sector consumes 54% of the world's total energy and 42% of India's electricity consumption [16]. For industrial applications, NILM methods face various hindrances like the need for high sampling data acquisition devices, difficulty in differentiating similar industrial loads, a small number of existing data sets for developing algorithms to disaggregate data [5], etc. But integrating smart and advanced energy monitoring solutions like NILM in industries will bring significant changes in energy efficiency.

10.8 WHY NILM IS NOT USED IN INDUSTRIES

- **Flexibility of models:** While residential/commercial buildings possess similar types of loads, for an industry, distinct and compatible hardware and software should be built based on the type of industry, the amount of equipment, and the type of equipment.
- **Hardware considerations:** The electrical meter must report data at a high frequency that is at least several data packets per minute or second-based transmission.
- **Complexity in disaggregation:** It is difficult to distinguish among industrial loads.
- **Nature of load:** Industries have dynamic loads, and hence high-frequency sampling is required.
- **Lack of technology alignment in industries**
- **Risk in privacy and security concerns for industrial data**

10.9 FUTURE SCOPE AND CONCLUSION

Several potential energy companies and research groups are working toward improving and commercializing NILM methods worldwide. Data science teams are constantly working on developing accurate load recognition algorithms [1]. The future implementation of NILM will reduce energy bills and the cost involved in individual submeter and sensor installations. With advancement in big data analytics and algorithms, NILM has a great scope in industrial applications, which will reduce energy losses and increase the reliability of industrial energy management.

REFERENCES

[1] S. Makonin and F. Popowich, "Nonintrusive load monitoring (NILM) performance evaluation: A unified approach for accuracy reporting," Energy Effic., vol. 8, no. 4, pp. 809–814, Jul. 2015. doi: 10.1007/s12053-014-9306-2.

[2] N. Batra et al., "NILMTK: An open-source toolkit for nonintrusive load monitoring," in e-Energy 2014 — Proceedings of the 5th ACM International Conference on Future Energy Systems, 2014, pp. 265–276. doi: 10.1145/2602044.2602051.

[3] J. Mei, S. Yan, G. Qu, D. He, and T. Habetler, "Design considerations for non-intrusive load monitoring hardware platforms in smart building," 2013. doi:10.1109/IAS.2013.6682581. https://ieeexplore.ieee.org/document/6682581

[4] S. Khan, A. F. Latif, and S. Sohaib, "Low-cost real-time nonintrusive appliance identification and controlling through machine learning algorithm," in 2018 International Symposium on Consumer Technologies, ISCT 2018, Jul. 2018, pp. 32–36. doi: 10.1109/ISCE.2018.8408911. https://ieeexplore.ieee.org/document/8408911

[5] M. Zhuang, M. Shahidehpour, and Z. Li, "An overview of nonintrusive load monitoring: Approaches, business applications, and challenges," in 2018 International Conference on Power System Technology, POWERCON 2018 — Proceedings, Jan. 2019, pp. 4291–4299. doi: 10.1109/POWERCON.2018.8601534.

[6] M. Figueiredo, A. de Almeida, and B. Ribeiro, "Home electrical signal disaggregation for non-intrusive load monitoring (NILM) systems," Neurocomputing, vol. 96, pp. 66–73, Nov. 2012. doi: 10.1016/j.neucom.2011.10.037.

[7] "Energy data intelligence for utilities | voltaware." https://voltaware.com/ (accessed May 13, 2021).

[8] "Technology—sense.com." https://sense.com/technology (accessed May 13, 2021).

[9] "Introduction to energy disaggregation." https://tech.ovoenergy.com/introduction-to-energydisaggregation/ (accessed May 13, 2021).

[10] "What is NILM technology? | Eco Efficient House." https://ecoefficienthouse.com.au/what-is-nilm/ (accessed May 13, 2021).

[11] H. Fu, "Review of lead-free halide perovskites as light-absorbers for photovoltaic applications: From materials to solar cells," Sol. Energy Mater. Sol. Cells, vol. 193, no. Nov. 2018, pp. 107–132, 2019. doi: 10.1016/j.solmat.2018.12.038.

[12] "Datacollection—Voltaware sensors." https://voltaware.com/our-technology/voltaware-sensors (accessed May 13, 2021).

[13] "Solution | Smart impulse." www.smartimpulse.com/en/solution/ (accessed May 13, 2021).

[14] "Benefits • meter track." www.metertrack.in/benefits/ (accessed May 13, 2021).

[15] "How does sense detect my devices?—Sense blog." https://blog.sense.com/articles/how-does-sense-detect-mydevices/ (accessed May 13, 2021).

[16] "Energy statistics." Accessed: May 13, 2021. [Online]. www.mospi.gov.in.

11 Energy Conservation Opportunities in an Educational Institution
Case Study at KPR Institutions

D. Mohankumar

CONTENTS

LEARNING OUTCOMES

At the end of this chapter, the reader will be able to understand:

- Energy conservation opportunities in colleges and large buildings.
- Energy saving calculations with technocommercial components.
- Real-time implemented savings for the energy saving opportunities identified.

DOI: 10.1201/9781003203810-11

11.1 ABOUT KPR INSTITUTIONS

KPR Institutions is an educational venture among KPR clusters, located in the eastern part of Coimbatore, Tamil Nadu, with lush green ambience and modern infrastructure. Located in the tropical part of India at 11'07" N and 77'14" E, it receives year-round direct solar radiation of 4 kWh/m². Having been built across 66.07 acres of land, the institution has a great potential to harvest solar energy to self-satisfy its energy needs. Though the campus is located in the industrial part of the city, it maintains green ambience and is a certified green campus.

The college has eight undergraduate programs and four postgraduate programs and has 3000+ students, 250+ faculties, and 100+ supporting staffs. As a part of renewable source effort, the institution has 13 kW of installed solar PV facility. The consumption achieves its peak during daytime when the machines at laboratories, computers, and accessories, along with mandatory equipment, are operating with heavy loads, in addition to the usage of the lights, fans, air conditioners. Without an Energy Audit, it would be tough to control or preserve electricity. The needs of the departments could be in jeopardy when a shortage of direct electricity might occur in the future if the conservation changeover doesn't occur now. We, the students of KPR Institute of Engineering and Technology, with full collaboration from our institution, conducted an Electrical Energy Audit, along with the identification of solar harvesting potential and other alternatives in order to suggest solutions for our institution should such a situation happen in near future.

For the case study, the power consumption of the institution from December 2019 to October 2020 was taken, and the monthly power consumption, including all the academic blocks, hostels, canteens, street lighting, sports complex, quarters, and other utilities, was noted. Figure 11.1 gives the details of consumption in the sample time period.

S. Number	Month' year	Units consumed (kWh)
1	December 2019	52,720
2	January 2020	61,144
3	February 2020	74,500
4	March 2020	59,412
5	April 2020	24,736
6	May 2020	30,412
7	June 2020	32,296
8	July 2020	33,100
9	August 2020	34,060
10	September 2020	34,204
11	October 2020	41,628

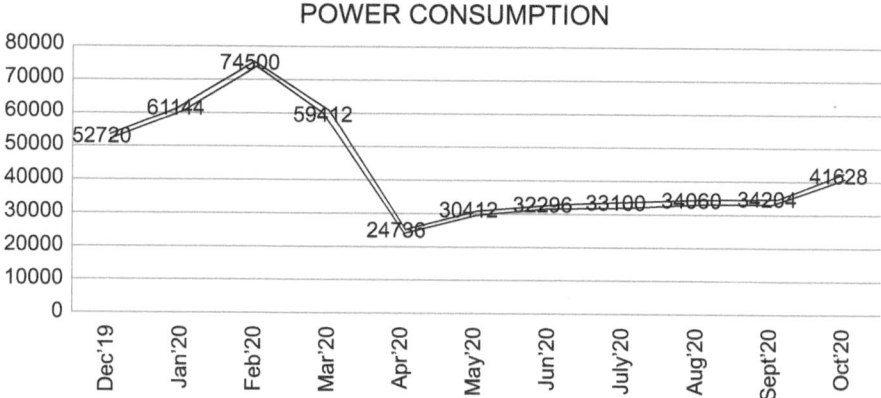

FIGURE 11.1   Power consumption pattern of the institution

In the trend shown, the peak consumption 74,500 kWh was observed to be consumed in the month of February 2020, which is a normal working month of the institution. The months of April to October are observed during the lockdown due to Covid-19 as having reduced consumption.

11.2 BUILDING CONSTRUCTION AREA DETAILS

The layout of the campus with a span of 2,70,009.321 m² is shown in Figure 11.2. The constructed area is also listed in Table 11.1 and shows the detailed area estimation of each floor in all blocks inside the campus. Upon adding all the concrete top area, the entire campus has more than approximately 5 acres of solar potential, and these 5 acres can support more than 1 MW solar power generation as per standard estimates. For research and laboratory purposes, the institute has already installed a 13 kW solar facility, which supports the grid power in meeting the institution's energy needs.

The grounds and street lighting system can also be changed to solar lighting systems. Playgrounds, indoor stadium, the gymnasium, parking facility, and other non-concrete areas can also be equipped with power from solar energy.

11.2.1 CAUSE LEAD TO AUDITING

The main source for the creation of electrical energy is thermal energy, through thermal power stations, with nonrenewable coal being one of the important fuels among the other sources. But the sources of coal are becoming limited due to the unrestricted exploitation by the mines of the planet. Now many major players are navigating toward alternate green forms like wind, hydro, solar, etc. Due to the main disadvantage of huge initial setup cost, our preference for them is very low, and now, with many subsidies from governments and improved efficiency,

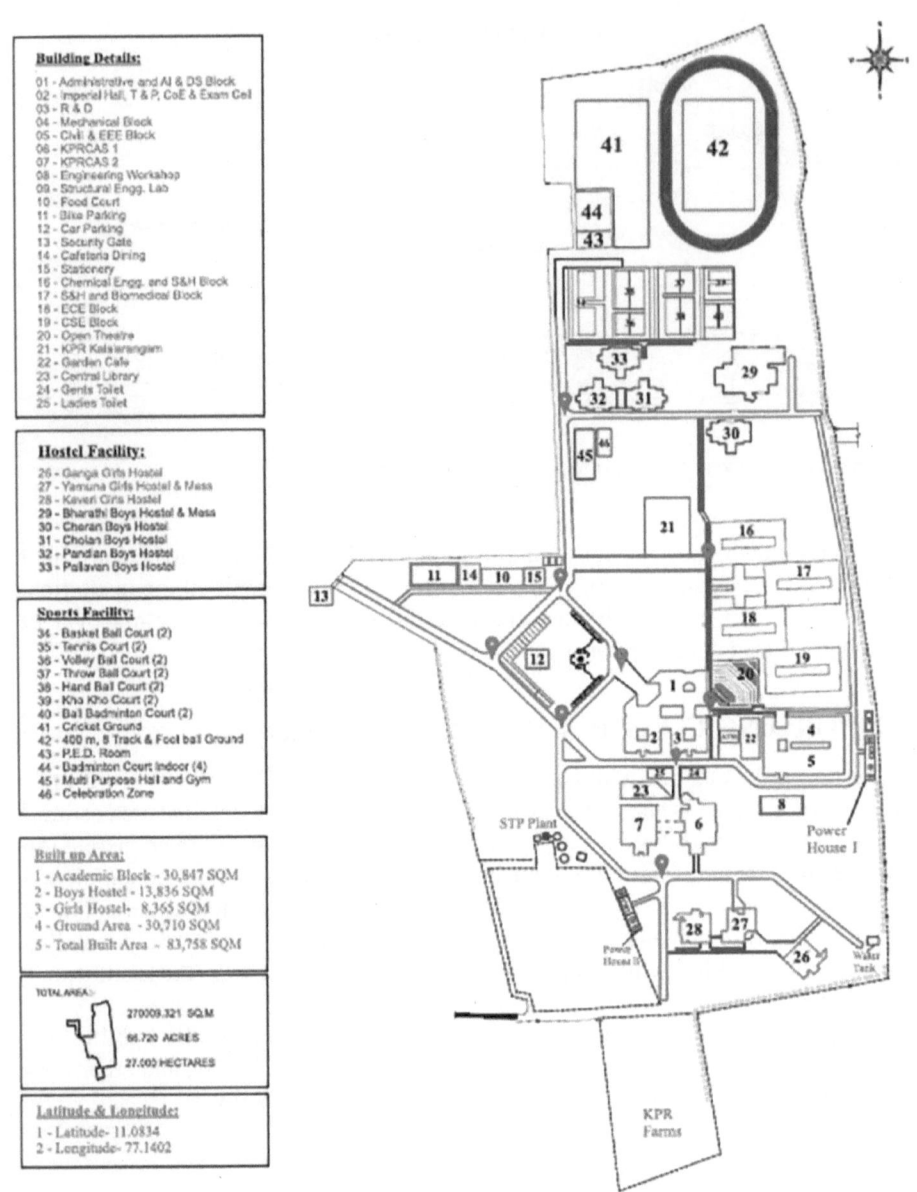

Building Details:

01 - Administrative and AI & DS Block
02 - Imperial Hall, T & P, CoE & Exam Cell
03 - R & D
04 - Mechanical Block
05 - Civil & EEE Block
06 - KPRCAS 1
07 - KPRCAS 2
08 - Engineering Workshop
09 - Structural Engg. Lab
10 - Food Court
11 - Bike Parking
12 - Car Parking
13 - Security Gate
14 - Cafeteria Dining
15 - Stationery
16 - Chemical Engg. and S&H Block
17 - S&H and Biomedical Block
18 - ECE Block
19 - CSE Block
20 - Open Theatre
21 - KPR Kalaiarangam
22 - Garden Cafe
23 - Central Library
24 - Gents Toilet
25 - Ladies Toilet

Hostel Facility:

26 - Ganga Girls Hostel
27 - Yamuna Girls Hostel & Mess
28 - Kaveri Girls Hostel
29 - Bharathi Boys Hostel & Mess
30 - Cheran Boys Hostel
31 - Cholan Boys Hostel
32 - Pandian Boys Hostel
33 - Pallavan Boys Hostel

Sports Facility:

34 - Basket Ball Court (2)
35 - Tennis Court (2)
36 - Volley Ball Court (2)
37 - Throw Ball Court (2)
38 - Hand Ball Court (2)
39 - Kho Kho Court (2)
40 - Ball Badminton Court (2)
41 - Cricket Ground
42 - 400 m, 8 Track & Foot ball Ground
43 - P.E.D. Room
44 - Badminton Court Indoor (4)
45 - Multi Purpose Hall and Gym
46 - Celebration Zone

Built up Area:

1 - Academic Block - 30,847 SQM
2 - Boys Hostel - 13,836 SQM
3 - Girls Hostel - 8,365 SQM
4 - Ground Area - 30,710 SQM
5 - Total Built Area - 83,758 SQM

TOTAL AREA:
270009.321 SQ.M.
66.720 ACRES
27.000 HECTARES

Latitude & Longitude:

1 - Latitude- 11.0834
2 - Longitude- 77.1402

FIGURE 11.2 Institute map with layout key.

TABLE 11.1

Campus Floor Area Estimation Sample.

No.	Area	Building	Floor	Area (ft2)
1	College Academic Block	ECE	GF	16,871
			FF	13,691
			SF	09,858
		Total		40,420
		EEE	GF	12,418
			FF	10,374
			SF	09,427
		Total		32,219
		Classroom block	GF	—
			FF	12,913
			SF	10,933
		Total		23,846
		Civil	GF	41,021
			FF	09,513
			SF	09,513
		Total		60,047
		Mech	GF	—
			FF	27,061
			SF	27,061
		Total		54,122
		CSE	GF	13,882
			FF	08,810
			SF	06,489
		Total		29,181
		S&H	GF	20,850
			FF	18,481
			SF	22,539
			TF	14,313
		Total		76,120
		Physics and chemistry lab	—	05,226
		Sports hall	—	06,257
		Workshop lab	—	05,227
		Structural lab	—	01,519
Total				332,665
2	Nonacademic block	Admin office	—	15,244
		Private canteen	—	08,222
		Entrance gate	—	00,511
		Bus shed	—	20,000
Total				43,977

the interest in the renewables increases. So obtaining electrical energy from thermal burning is still being used, even with the utmost efficient in continuous usage, and we need to conserve electricity in order to reduce contributions to greenhouse gas emissions.

11.3 NEED FOR ENERGY AUDIT

The Energy Audit is the preliminary step to reduce energy costs and preserve natural sources. The Audit helps us find areas where energy savings can be made, work out which changes should be made first, and understand the costs and benefits of implementing these changes. Assessments are best carried out as part of an ongoing energy management program. Behavioral change through staff education can often provide the greatest benefit at the lowest cost. Even small savings across one or more areas can achieve a large saving overall, and so the audit must review both the detailed and the general energy use practices. The physical initiatives already underway have sparked many others, but due to lack of ready ideas, those initiatives slowly evaporate. So now we need some method for preparing the ideas of conservation. Toward that end, the method of *auditing* is recommended. As in financial auditing, for the economic stability of the institution, the Energy Audit will also reduce the financial costs of the electricity, which in turn will pave the way for the economic stability of the institution and for the natural stability of our Mother Earth.

11.3.1 Additional Details

Place	Area of Each (ft2)	Lumens
Classroom	1220.00	Along corridor: 1704 Inside corridor: 2370 (having windows outside of them)
Department Staff Room	4859.90	Without light: 303 With light: 2128
Core Laboratory	2065.70	Without light: 33 With light: 742
Seminar Hall	2852.52	Without light: 0 With light: 2304
Office	2152.78	Without light: 700 With light: 3000

For the classrooms and the office, the consumption for lighting is mostly avoided, as they are naturally luminated by sunlight. So their lumens with light are not taken. Seminar Hall without light is 0 lux. So for that with light alone considered.

11.4 PLAN OF ACTION

S. No.	Phase of Audit	Purpose and Result
Step 1	**Phase I: preaudit phase** • Plan and organize team for audit • Conduct a brief meeting with others to create awareness and discuss • About energy audit	• Establish energy audit team and organize instruments • Building up cooperation among team members • Awareness creation and getting a clear picture of what to do
Step 2	**Phase II: audit phase** • Gather primary data for energy audit • Analysis of the use of energy • Analysis of the cost–benefit • Create a report and presentation on this	• Analysis of historic data • Collect data of annual energy bill and energy consumption for each month • Analysis of all the data and brain streaming with members on energy loss • Analysis of the cost that is benefitted from this • Create a report and presentation to management
Step 3	**Phase III: postaudit phase** • Implement the action plan and follow that plan	• Monitor performance • Implementation of plan, follow up, and review it

11.5 CONSUMPTION DETAILS

The consumption rates are calculated based on the tarriff rates of Tangedco for Tamil Nadu for the active months of academics. For citations on the tarriff rates of the Private Educational Institutions (II B) [3], see the reference.

From the tarriff rates for Private Educational Institutions, we calculated with the amount of Rs. 6.35 for a single unit, with the demand of 350 Rs./KVA/month.

11.5.1 Mech and Civil Block

Components	Wattage	Count	Total capaciy in kW	Hours per day	Units in kWh for daily usage	Monthly usage in kWh	Tarriff rate per month
Fluorescent lights	40	425	17	4	68	0272.00	01,727.20
Led lights	15	128	1.92	3	5.76	0017.28	00,109.72
CFL lamps	15	60	0.9	8	7.2	0216.00	01,371.60
Fans	53	268	14.2	6	852	2556.00	16,230.60
Systems in designing cad labs	162	266	44.15	3.6	158.94	4768.00	30,278.07
Laptops	60 W	50	03.00	2	006.00	0180.00	01,143.00
Projectors	282	6	1.6	1	1.6	0048.00	00,304.80
Printers	240	5	1.2	0.34	0.408	0012.24	00,078.00
AC in Dhanam Hall	50 tonnage OR 14.2 kW	1 centralized air conditioner	14.02	3	042.06	01216.8	08,012.43
Air conditioner	281.31	20 T	5.63	4	022.52	0675.60	04,920.06
All core laboratories			188.11	20.4	173.03	5190.90	32,962.21
Grand total			290.68 kW	69.74	556.85	16705.5	Rs. 1,06,079.92

11.5.2 EEE Block

Components	Type	Wattage	Quantity	Capacity (kW)	No. of hours per day	Units (kWh)	Monthly usage (kWh)	Tarriff per month
Fluorescent lights	Crompton—FTL Super Gold	40	146	05.84	4	23.36	0700.80	04,450.08
Fans	Crompton Greaves—Riveria	53	120	06.36	6	38.16	1144.80	07,269.48
Printers	Canon 2900	240	4	00.96	0.34	00.32	0009.60	00,060.96
Desktops	17 inch Dell TFT	162	5	00.81	4	03.24	0097.20	00,617.22
LED TV	Onida 32 HIF	40	1	00.40	0.34	00.13	0003.90	00,024.76
Projectors	Benq	282	4	01.12	1	01.12	0033.60	00,213.36
Faculty While Charging Their Laptops	Mainly Dell and HP	60	50	03.00	2	006.00	0180.00	01,143.00
415 V Motor	Sample lab test equipment	2982	9	26.83	1	26.83	0791.4	05,025.39
220 V Motor	Sample lab test equipment	3728	5	18.64	1	18.64	0559.20	03,550.92
DC lab 220 V motor	Sample lab test equipment	2237	12	26.84	1	26.84	0805.20	05,113.02
Simulation center and other designing centers								
Desktops	17 inch Dell TFT	162	120	19.44	2	38.88	1166.40	07,406.64
Air conditioners	Mitusubishi	281.86	10	02.81	2	05.62	0168.60	01,070.61
Grand total for the EEE block		10177 W	486	113.05 kW	24.68	189.14 kWh	5674.20 kWh	Rs. 36,031.17

11.5.3 Admin Block

Components	Type	Wattage	Quantity	Capacity (kW)	Hours per day	Units (kWh)	Monthly (kwh)	Tarriff per month
LED lights	Crompton—LLD20CDRL	20	197	03.94	2	007.88	0236.40	01,501.14
CFL lamps	Crompton Greaves—Riveria	53	16	00.84	8	006.72	0201.60	01,280.16
Printer	Canon 2900	240	4	00.96	1	000.96	0028.80	00,182.88
Desktop	17 inch Dell TFT	162	20	03.24	7	022.68	0680.40	04,320.54
LED TV	Onida 32 HIF	55	7	00.38	1	000.38	0011.40	00,072.39
Projector	Benq	282	2	00.56	1	000.56	0016.80	00,106.68
Air conditioners	SPLIT AC Mitsubishi	14068.0	10	14.06	8	112.54k	3376.32	21,439.63
Air conditioners	Centralized Mitsubishi 65 Tonnage	18288.4	1	18.28	6	109.68	3290.40	20,894.04
Grand total for admin block		33178.4 W	261	42.26 kW	32	261.40 kWh	7842.00 kWh	Rs. 49,796.70

11.6 CONSERVATION SUGGESTIONS

Building Energy Management Systems (BEMS) is a dedicated computer and network that controls and monitors all the equipment (such as pumps, fans, dampers, chillers, lighting, renewable energy system, elevators, etc.) that are part of the building HVAC system. They can provide very sophisticated control, but their influence on energy efficiency depends on how they are designed. BEMS control based on energy efficiency rather than on temperature control can realize substantial energy savings. As sensor prices have come down, installing more sensors enables more sophisticated approaches to monitor and control the system more effectively.

11.6.1 LIGHTING

From the preceding tabulations, we can see that the use of lights and fans are by themselves very high, when compared to the other basic amenities. So we need to reduce the lighting facility to the required need, but, instead of reduction, we can adopt other types of light that consume less power. Until the Mech, Civil & EEE departments need to change their lighting from ordinary fluorescent lights to the new LED type of conventional lights for the classrooms, cabins, labs, etc. In this case, we can change the fluorescent lights out for LED tubes. For example, a product from Philips, a 121 cm tube having a luminous flux of 2000, will be adequate for a classroom. It costs less with Rs.1067 for a pack of four, has a warranty period of a year, and may have an extended life span of a 3–4 years. It consumes 20 W.

11.6.2 FANS

Fans are frequently used amenities in the college. They are used throughout all the academic hours, whereas lights are used only when it is dark, and during the night hours, only the corridor lights are used. So we have to conserve the energy used by fans. Energy efficient 32 W fans with a warranty period of 2 years have been suggested. At low speeds, it has only a 7 W consumption. Also, it has an extended life span of up to 5–6 years.

11.6.3 PROJECTORS, PRINTERS, DESKTOPS

We appreciate the college's selection of low-power-consuming digital education tools and its conserving of energy consumption. We request that the college follow the same brands as they are using now with the same working hours.

11.6.4 AIR CONDITIONERS

As mentioned in the earlier tables, most of the class rooms and seminar halls are equipped with centralized air conditioning systems. It has been suggested to clean the ducts periodically to avoid the settling of dust which can resist the air flow and cause inefficiency in the system.

11.7 RESULTS AND DISCUSSION

For the change of tariff rates of lighting, see Figure 11.3. For the change of tariff rates for fans, see Figure 11.4. Based on these figures, we making the suggestions mentioned in "Conservation Suggestions." If we follow those conservations and update with new modern technologies, we can definitely reduce power consumption, save the nature, and ensure the economic stability of the institution and mainly the natural stability of our land.

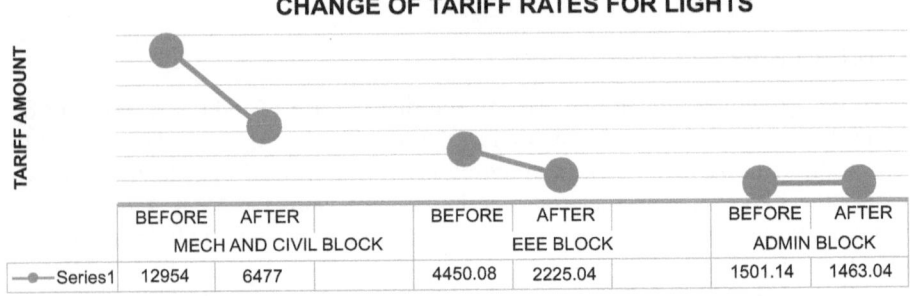

	BEFORE	AFTER		BEFORE	AFTER		BEFORE	AFTER
	MECH AND CIVIL BLOCK			EEE BLOCK			ADMIN BLOCK	
—●—Series1	12954	6477		4450.08	2225.04		1501.14	1463.04

FIGURE 11.3   Change of tariff rates for lights after LED lights installation.

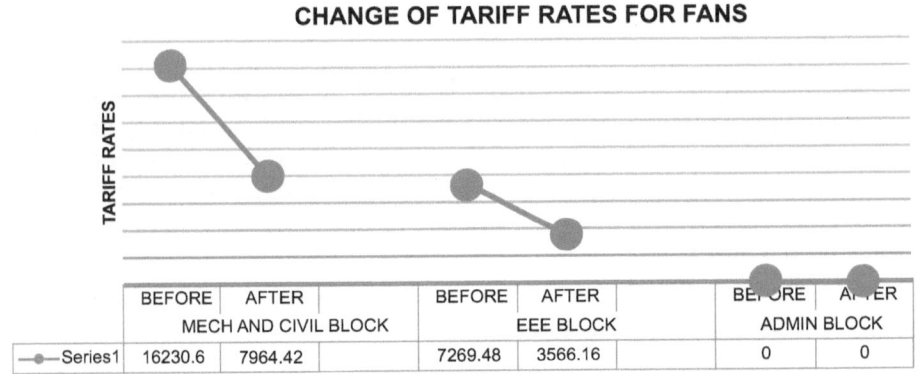

	BEFORE	AFTER		BEFORE	AFTER		BEFORE	AFTER
	MECH AND CIVIL BLOCK			EEE BLOCK			ADMIN BLOCK	
—●—Series1	16230.6	7964.42		7269.48	3566.16		0	0

FIGURE 11.4   Change of tariff rates for lights after EE fans installation.

11.8 PAYBACK PERIOD

Normal consumption of lights and fans Rs. 1,68,218.00 (a)
Conservation consumption of lights and fans Rs. 67,192.00 (b)
Difference on energy billing savings/month Rs. 1,01,026.00/month (a – b)

New replacement of lights and fans considering Conservation Suggestion:

Replacement cost of lights = no. of lights × price
                           = 732 × 247.25
                           = Rs. 1,80,987
Replacement cost of fans   = no. of fans × price
                           = 1189 × 3000
                           = Rs. 35,67,000
Total replacement cost   = Rs. 37,47,987

Therefore, payback period = Total replacement cost/savings per month
                          = 37,47,987/1,01,026
                          = 37.09 months
                          = 3.09 years

From these calculations, we can determine that after a little less than 5.5 years, our investment will be recovered, saving nature.

11.9 GENERAL RECOMMENDATIONS

1. Replace lights and fans with energy efficient ones.
2. Use smart power strips, which eliminate power loss during standby mode.
3. Replace lights and fans with new ones.
4. Switch off lights and fans when they are not in use.
5. Try to make use of natural light, which can reduce usage of lights during daytime.
6. Clean the blades of fans and the surfaces of lights of dust on a regular basis, which improves the efficiency of the fans and the luminous intensity of lights.
7. The installed solar setups can be multiplied as far as possible, and frequent maintenance of them required.
8. Electrical connections should be checked for current discharge.
9. Use of LED lights instead of incandescent lights will consume less power, and luminosity will be high.
10. Frequent suggestions should be encouraged among the beneficiaries or the functionaries of the institutions (e.g., students in the college) in order to improve on the effects of the changes.

11.10 BUILDING SETUP

Daylighting: Daylighting is the controlled admission of natural light, direct sunlight, and diffused skylight into a building to reduce the use of electric lighting and saving energy. The appropriate use of windows, skylights, and other apertures in the building helps to harvest daylight. The various methods include external light shelves, light pipes (for large window area), skylights, and roof monitors (for areas without access to windows), and light-colored interior surfaces that reduce luminance contrast and improve coverage. Large buildings can allow daylight into more spaces by having central courtyards or atria or by having other cutouts in the building form.

Light shelves: A light shelf is a passive architectural device used to reflect natural daylight into a building. Unlike direct sunlight, which can cause glare near an opening and leave dark spaces further inside the room, the light shelf bounces sunlight off a horizontal surface and distributes it more evenly and deeply within a space. The light shelf is often designed as part of a broader daylight and shading strategy.

Chilled water storage: Chilled water storage allows chillers to operate at times of day that differ from when air conditioning is needed. In a chilled water system, chilled water is typically created overnight (when chillers operate more efficiently due to the cooler ambient temperature), is stored in very large well insulated tanks (designed to allow the coldest water to sink and warmer water to rise), and is drawn upon as needed. In locations with a strong so-called diurnal swing in temperature, the additional energy for pumping is more than offset by the lower chiller energy.

Ice bank: This system is similar to a chilled water system. Blocks of ice are created at night during off-peak periods, typically nighttime. The chiller cools an ethylene glycol solution to below 0°C, and the solution is circulated through tubes in a tank, freezing the water held in the tank. During the day, the ice melts, cooling the solution in the tubes. The chilled solution is moved through a heat exchange coils where it cools the air.

Electronically commutated fans: Electronically commutated (EC) fans use brushless motors with permanent magnets and DC voltage controlled by a microprocessor like the fans found in desktop computers. These motors are more energy efficient than conventional AC motors because they do not have the same copper wire windings. The speed of EC fans can be controlled without the need for an external variable frequency drive (VFD). This measure is applicable for fans throughout the building (for example, in fan coil units)

11.11 CONCLUSION

Considering the data collected, the basic amenities of projectors, printers, desktops in designing centers, and generally the equipment in the laboratories are used very wisely. It is recommended to keep doing the same. The unnecessary wastage of power through lights and fans is very obvious. Alternatives have been given, and the reduction in power consumption is shown for reference. We also recommend proper care for the plants at the institute that are growing into trees and bringing natural ventilation throughout the blocks. If any alterations are done in the future, we recommend recording the lumens of light intensity, ventilation of air, and radiance of the solar light and then executing the resultant plan. Though the use of the equipment for such measurements is minimal, the need is great, leading to sound planning of building structures. The select buying and maintenance of those instruments are recommended, so that they help in the execution of plans. It is suggested to have a solar panel above the Mech & Civil block, as in the EEE block, or the connection of that power to the Mech & Civil block. Finally, we recommend a yearly energy audit for the institution, which will clarify usage by the institution.

12 Optimization Techniques for Optimal Power Flow

A Precursor

L. Chitra

CONTENTS

LEARNING OUTCOMES

At the end of this chapter, the reader will be able to understand:

- Optimal power flow techniques (OPF).
- The need for optimal power flow optimization.
- Optimization methods for OPF.

12.1 INTRODUCTION

Today's power industries are moving toward a very competitive market which greatly influences the entire operation. Economical and security issues of such systems are highly coordinated, warranting the need for optimizing tools in supporting system control and management [1]. The optimal power flow (OPF) is considered an important tool in power system analysis owing to its capacity in dealing with different situations. It is the backbone application of any power system [2]. The OPF problem has been studied by many researchers since 1962 [3]. OPF optimization is a mathematical approach in solving the global power system problem, which is aimed at computing fewer control movements in keeping power systems at a desirable state [4]. OPF is extensively employed in various applications, namely voltage control and constrained economic dispatch problems [5]. When applied to large networks, OPF poses greater complexity, limiting its usage to such networks. [6]. The OPF problem aims to achieve an optimal solution of a specific objective function (power system), like fuel cost, by adjusting

DOI: 10.1201/9781003203810-12

the control variables and simultaneously satisfy physical and operational constraints [7]. Some traditional optimization methods in solving OPF employ gradient methods, quadratic programming methods, etc. A holistic survey of such methods is presented in [8] and [9]. As stated earlier, some of the shortcomings in the application of these methods in industrial sector are low practicability (theoretical assumption), inability to handle integer or binary variables, etc. Recently, many stochastic and heuristic optimization techniques have been developed and applied to global optimization problems. These methods offered excellent global search ability over traditional algorithms [10]. Many bioinspired algorithms, such as ant colony optimization (ACO), bacterial foraging algorithms (BFA), genetic algorithms (GA), differential evolution (DE), particle swarm optimization (PSO), search algorithms like tabu, and gravitational, were widely used in solving the OPF. They have their own merits and demerits in attaining global optima. A review of these algorithms was reported in [11]–[13].

12.2 OVERVIEW OF THE OPF PROBLEM

Researchers have used different scenarios to handle the problem of OPF using different strategies contributing to power system analysis. For example, OPF can be employed to reduce total real power loss through reactive power dispatch. For power system planning and operation, optimal power flow acts as an important tool. The optimal power flow consists of the objective function within the balanced and unbalanced constraints. In planning and operation requirements, the generation cost is minimized by adjusting the generated powers of the requirement. So the input powers and bus voltages are assumed in such a way that the minimum operating cost will be achieved. The voltages at the generator nodes along with the input powers has to be determined at this point. The choices for the input degree of freedom may be largely due to cost minimization.

12.3 OPTIMIZATION METHODS FOR OPF

This section outlines various optimization methods that aim to solve the OPF problem. Defining multiple objectives is done in solving this problem. Bioinspired algorithms, evolutionary computing algorithms, swarm-based methods, etc. have shown promising results overcoming the shortcomings of traditional methods [14]–[18]. Some of the methods are discussed here.

12.3.1 GENETIC ALGORITHM (GA)

Genetic algorithms [19] are stochastic search algorithms simulating the biological evolution process. GA was first employed in power systems for load flow problems. The GA was able to compute the normal load flow solution for small-scale networks by modifying the cost function. Authors in [20] investigated various approaches for improving the convergence rate of GA. A generalized OPF aims to minimize the costs of meeting a power system's load demand without compromising the security aspect. Each system must be operated within its safe operating limit, which includes minimum and maximum outputs for generators, megavolt-ampere (MVA) flows on transformers and distribution lines, with controlled system bus voltage. Bakritzs [21] reported the application of GA to the economic dispatch problem. GA was able to run on parallel machines where there was no restriction on the generator objective function and performed better than dynamic programming methods. However, the method had the drawback of execution time apparently increasing with system size. Some other modifications were also made to GA in applying for OPF to enhance convergence [22]–[23]. The OPF problem as encountered was multiobjective in which there was the simultaneous optimization of control parameters. This included reduction in fuel cost, fuel loss, voltage stability index, etc.

12.3.2 PARTICLE SWARM OPTIMIZATION (PSO) ALGORITHM

A metaheuristic population-based algorithm developed by Kennedy and Eberhart [24] was employed to solve the OPF problem [25]. Owing to the merits of PSO, researchers have extensively applied it in solving optimization problems in electric power systems and have achieved success in specific areas [26].

The reconstruction operator was fused with PSO in solving the OPF problem, with security constraints [27].

Initially, the weight was defined and made to vary as the process proceeds in finding a global search solution. It was an iterative process that yielded a solution that satisfied generation and operative constraints. Harinder Pal Singh et al. [28] proposed a similar technique using PSO and applied on the IEEE 9-bus system with objectives like fuel cost, transmission loss, etc. Attractive results proved the system's validity and efficiency. The method attained a reduction in voltage deviation by 11.2% and a 23% reduction in transmission loss with a rise of 10% in fuel cost when compared with different scenarios. Using PSO, the authors in [29] proved that the IEEE 30-bus system was capable of generating an optimal power of 193 MW. Singh et al. [30] developed an aging leader and challengers PSO (ALC-PSO) for solving various OPF types with multiple complexities.

12.3.3 GLOWWORM SWARM OPTIMIZATION (GSO) ALGORITHM

A multiobjective OPF was proposed by [31] in which the OPF problem was solved using the cost of generation and by reducing emissions as cost functions. This method avoided the compromise between the control parameters in obtaining an optimal power flow in large-scale sectors. The robustness and efficiency of the proposed model were tested on the IEEE 30-bus and practical Indian 75-bus systems for cost minimization as an objective function, and the IEEE 30-bus test system for minimization of cost and emission as objectives. The outcomes arrived at from both the PSO and GSO were compared on the basis of various control parameters

12.3.4 SEARCH AND HYBRID ALGORITHMS

A differential search algorithm was reported by Abaci and Yamacli [32] in solving OPF with multiple constraints. Backtracking the search algorithm (BSA) was used by Ayan and Kiliç [33] to find a solution for the OPF problem, and it was applied to biterminal high-voltage direct current (HVDC) power systems. Combining the benefits of the firefly algorithm (FFA) and pattern search (PS) algorithm, a hybrid approach for a power system OPF problem was presented by Mahdad and Srairi [34] to reduce fuel cost, power loss, and voltage deviation. A novel and effective hybrid algorithm combining firefly and particle swarm optimization (HFPSO) algorithm was applied in solving different convex and nonlinear OPF problems [35]. The novel combination was developed with an approach to improve the rate of exploration and exploitation and also to obtain a faster rate of convergence. Different cost functions were studies including reduction of generation cost, improvement of voltage profile and voltage stability, reduction of power loss during transmission, etc. MATLAB was used as a platform to test the model using the IEEE 30-bus system. On comparison with standard techniques, the method was able to create optimal solutions with faster convergence and solve complicated OPF problems.

A fuzzy-based hybrid PSO was developed by [36]. This method concentrated on load demand and speed of the wind for forecasting and on avoiding errors. These parameters were annotated using fuzzy sets, and membership functions were derived under an uncertainty ambience. IEEE 30- and IEEE 118-bus test systems were used to validate their model. Using these functions, PSO was incorporated with a local random search in reaching the optimal set for the real generation of power levels, voltage magnitudes, and LTC tap positions under fuzzy ambience.

Optimization of OPF was addressed using the hybridization of particle swarm optimization and gravitational search algorithm (GSA) [37]. The IEEE 30- and IEEE 118-bus test systems were used to validate their model. The objectives included minimization in fuel cost, enhancement in voltage profile and voltage stability, reduction of power loss, Etc. by considering the valve point of generation. A similar approach was developed by Attia et al. [38] for single- and multiple-objective OPF optimization by combining grey wolf optimizer (GWO) and differential evolution (DE) algorithms. An integrated an approach using invasive weed optimization (IWO) and Powell's pattern search (PPS) was proposed by Kaur and Narang [39]. The solution obtained by the IWO was improved by the PPS algorithm as applied to AC transmission system devices for the OPF problem. The system was tested in IEEE bus systems, which reduced sequentially three cost functions: fuel cost, pollutant emission, and system transmission loss. A T-test performed proved the statistical performance of the method comparatively.

12.4 CONCLUSION

Optimal power flow is regarded as an effective tool for optimization in power system operation scheduling, analysis, and management of energy. The OPF problem consists of a cost function optimization in multiple forms that satisfies a set of physical and operational constraints. Many researchers have made efforts in solving this problem over the decades. Hence a comprehensive precursor involving various optimizations via bioinspired, hybrid, and search methods is presented. Only a few of those are presented taking into consideration different cases and scenarios in which the research is carried on. New algorithms involving metaheuristic searches such as GSO, PSO, and GSA have the capacity to handle integer variables extremely well but warrant deeper investigation with respect to performance under various multiple systems. Though the literature shows promising results, the capacity of these algorithms in providing satisfying solutions in the case of large-scale power systems is yet to gain the attention and be well demonstrated. A hybrid algorithm will aid researchers in comparing and selecting a valid and appropriate OPF method in finding an optimal state of any power system involving single or multiple constraints.

REFERENCES

[1] Wu F, Tong X, Zhang Y, A decoupled semismooth newton method for optimal power flow. In IEEE Power Engineering Society General Meeting, 18–22 June 2006, 2006.

[2] Squires R B, Economic dispatch of generation directly from power system voltages and admittances. Transactions of the American Institute of Electrical Engineers. Part III: Power Apparatus and Systems, vol. 79, no. 3, pp. 1235–1244, April 1960. doi: 10.1109/AIEEPAS.1960.4500947.

[3] He S et al, An improved particle swarm optimization for optimal power flow. In International Conference on Power System Technology, 21–24 November 2004, pp. 1633–1637, 2004.

[4] Azmy A M. Optimal power flow to manage voltage profiles in interconnected networks using expert systems. IEEE Transactions on Power Systems, vol. 22, pp. 1622–1628, 2004.

[5] Yesuratnam G, Thukaram D, Fuzzy—expert approach for voltage-reactive power dispatch. In IEEE Power India Conference, 10–12 April 2006, pp. 8, 2006.

[6] Azmy A M, Optimal power flow to manage voltage profiles in interconnected networks using expert systems. IEEE Transactions on Power Systems, vol. 22, pp. 1622–1628, 2004.

[7] Chan K W, Pong T Y G, Mo N, Zou Z Y, Transient stability constrained optimal power flow using particle swarm optimisation. IET Generation, Transmission and Distribution, vol. 1, no. 3, pp. 476–483, 2007.

[8] Huneault M, Galiana F D, A survey of the optimal power flow literature. IEEE Transactions on Power Systems, vol. 6, no. 2, pp. 762–770, May 1991. doi: 10.1109/59.76723.

[9] Frank S, Steponavice I, Rebennack S, Optimal power flow: A bibliographic survey I. Energy Syst, vol. 3, no. 221–258, 2012. doi: 10.1007/s12667-012-0056-y

[10] Lambert-Torres G, Esmin A A A, Loss power minimization using particle swarm optimization. In International Joint Conference on Neural Networks, 2006. IJCNN '06

[11] AlRashidi M R, El-Hawary M E, Applications of computational intelligence techniques for solving the revived optimal power flow problem. Electric Power Systems Research, vol. 79, no. 4, pp. 694–702, 2009. doi: 10.1016/j.epsr.2008.10.004

[12] Kalia Chirag, Sharma Bhupender Artificial intelligence techniques for optimal power flow. International Journal of Engineering Research and Technology, vol. 1, no. 2, 2013.

[13] Barbulescu C, Kilyeni S, Simo A, Oros C, Artificial intelligence techniques for optimal power flow. In: Balas V., Jain L., Kovačević B. (eds) Soft Computing Applications. Advances in Intelligent Systems and Computing, vol. 357. Springer, Cham, 2016. doi: 10.1007/978-3-319-18416-6_101

[14] Lee K Y, Vlachogiannis J G, Fuzzy logic controlled particle swarm for reactive power optimization considering voltage stability. In The 7th International Power Engineering Conference, 2005. IPEC 2005, pp. 1–555, 2005.

[15] Lee K Y, Vlachogiannis J G, Hatziargyriou N D, Ant colony system-based algorithm for constrained load flow problem. IEEE Transactions on Power Systems, vol. 20, no. 3, pp. 1241–1249, 2005.

[16] Meng F, Zhang H, Zhang L, Reactive power optimization based on genetic algorithm. In International Conference on Power System Technology, 1998. Proceedings. POWERCON apos; 98, vol. 2, pp. 1448–1453, August 1998.

[17] Chuanwen J, Zhiqiang Y, Zhijian H. Economic dispatch and optimal power flow based on chaotic optimization. In International Conference on Power System Technology, 2005. Proceedings. PowerCon 2002, vol. 4, pp. 2313–2317, 2002.

[18] Ma J T, Wu Q H, Power system optimal reactive power dispatch using evolutionary programming. IEEE Transactions on Power Systems, vol. 10, no. 3, pp. 1243–1249, 1995.

[19] Walters D, Sheble G B, Genetic algorithm solution of economic dispatch with valve point loading, IEEE Trans. on Power Systems, vol 8, no. 3, pp. 1325–1331, 1993.

[20] Cao Y J, Zhao B, Guo C X, An improved particle swarm optimization algorithm for optimal reactive power dispatch. In IEEE Power Engineering Society General Meeting, 12–16 June 2005, vol. 1, pp. 272–279, June 2005.

[21] Bakritzs A, Perirtridis V, Kazarlis S, Genetic algorithm solution to the economic dispatch problem, IEE Proc.,-Generation Transmission Distribution, vol. 141, no.4, pp. 377–382, July 1994.

[22] Kumari MS, Maheswarapu S, Enhanced genetic algorithm based computation technique for multiobjective optimal power flow solution, International Journal of Electrical Power and Energy Systems, 2010. https://doi.org/10.1016/j.ijepes.2010.01.010

[23] Attia AF, Al-Turki YA, Abusorrah AM, Optimal power flow using adapted genetic algorithm with adjusting population size. Electric Power Components and Systems, 2012. https://doi.org/10.1080/15325008.2012.689417

[24] Kennedy J, Eberhart R. Particle swarm optimization, Proceedings of IEEE International Conference on Neural Networks, vol. IV, pp. 1942–1948. Neural Networks. 1995.

[25] Abido MA. Optimal power flow using particle swarm optimization, International Journal of Electrical Power Energy Systems, 2002. https://doi.org/10.1016/S0142-0615(01)00067-9

[26] Zhao Z, Yang B, Chen Y. Survey on applications of particle swarm optimization in electric power systems. In IEEE International Conference on Control and Automation, May 30, 2007—June 1, 2007. ICCA 2007, pp. 481–486, 2007.

[27] Ramirez J M, Onate P E. Optimal power flow solution with security constraints by a modified PSO. In IEEE Power Engineering Society General Meeting, 24–28 June 2007, pp. 1–6, 2007.

[28] Singh H P, Brar Y S, Kothari D P, Optimal power flow solution using particle swarm optimization, International Journal of Computational Engineering Research, vol. 8, no. 11, pp. 33–29, 2018.

[29] K Widarsono et al. 2020 J. Phys.: Conf. Ser. 1595 012033.

[30] Singh R P, Mukherjee V, Ghoshal S P, Particle swarm optimization with an aging leader and challengers algorithm for the solution of optimal power flow problem, Applied Soft Computing, vol. 40, pp. 161–177, 2016.

[31] Reddy S S, Rathnam C S, Optimal power flow using glowworm swarm optimization, International Journal of Electrical Power & Energy Systems, vol. 80, pp. 128–139, 2016. doi: 10.1016/j.ijepes.2016.01.036

[32] Abaci K, Yamacli V, Differential search algorithm for solving multi-objective optimal power flow problem, International Journal of Electrical Power & Energy Systems, vol. 79, pp. 1–10, 2016.

[33] Ayan K, Kiliç U, Optimal power flow of two-terminal HVDC systems using backtracking search algorithm, International Journal of Electrical Power & Energy Systems, vol. 78, pp. 326–335, 2016.

[34] Mahdad B, Srairi K, Security optimal power flow considering loading margin stability using hybrid FFA-PS assisted with brainstorming rules, Applied Soft Computing, vol. 35, article no. 3031, pp. 291–309, 2015.

[35] Khan A, Hizam H, bin Abdul Wahab NI, Lutfi Othman M, Optimal power flow using hybrid firefly and particle swarm optimization algorithm. PLoSONE, vol. 15, no. 8, e0235668, 2020. doi: 10.1371/journal.pone.0235668

[36] Liang Ruey-Hsun, Tsai Sheng-Ren, Chen Yie-Tone, Tseng Wan-Tsun, Optimal power flow by a fuzzy based hybrid particle swarm optimization approach, Electric Power Systems Research, vol. 81, no. 7, pp. 1466–1474, 2011, doi: 10.1016/j.epsr.2011.02.011

[37] Radosavljević Jordan, Klimenta Dardan, Jevtić Miroljub, Arsić Nebojša, Optimal power flow using a hybrid optimization algorithm of particle swarm optimization and gravitational search algorithm, Electric Power Components and Systems, vol. 43, no. 17, pp. 1958–1970, 2015. doi: 10.1080/15325008.2015.1061620

[38] El-Fergany Attia A, Hasanien Hany M, Single and multi-objective optimal power flow using grey wolf optimizer and differential evolution algorithms, Electric Power Components and Systems, vol. 43, no. 13, pp. 1548–1559, 2015. doi: 10.1080/15325008.2015.1041625

[39] Kaur M, Narang N. An integrated optimization technique for optimal power flow solution. Soft Comput, vol. 24, pp. 10865–10882, 2020. doi: 10.1007/s00500-019-04590-3